Functional and Smart Materials

Manufacturing Design and Technology Series

Series Editor:
J. Paulo Davim

This series will publish high-quality references and advanced textbooks in the broad area of manufacturing design and technology, with a special focus on sustainability in manufacturing. Books in the series should find a balance between academic research and industrial application. This series targets academics and practicing engineers working on topics in materials science, mechanical engineering, industrial engineering, systems engineering, and environmental engineering as related to manufacturing systems, as well as professions in manufacturing design.

Drills
Science and Technology of Advanced Operations
Viktor P. Astakhov

Technological Challenges and Management
Matching Human and Business Needs
Edited by Carolina Machado and J. Paulo Davim

Advanced Machining Processes
Innovative Modeling Techniques
Edited by Angelos P. Markopoulos and J. Paulo Davim

Management and Technological Challenges in the Digital Age
Edited by Pedro Novo Melo and Carolina Machado

Machining of Light Alloys
Aluminum, Titanium, and Magnesium
Edited by Diego Carou and J. Paulo Davim

Additive Manufacturing
Applications and Innovations
Edited by Rupinder Singh and J. Paulo Davim

For more information about this series, please visit: https://www.routledge.com/Manufacturing-Design-and-Technology/book-series/CRCMANDESTEC

Functional and Smart Materials

Edited by
Chander Prakash, Sunpreet Singh, and
J. Paulo Davim

CRC Press is an imprint of the
Taylor & Francis Group, an **informa** business

First edition published 2021
by CRC Press
6000 Broken Sound Parkway NW, Suite 300, Boca Raton, FL 33487-2742

and by CRC Press
2 Park Square, Milton Park, Abingdon, Oxon, OX14 4RN

© 2021 Taylor & Francis Group, LLC

CRC Press is an imprint of Taylor & Francis Group, LLC

Reasonable efforts have been made to publish reliable data and information, but the author and publisher cannot assume responsibility for the validity of all materials or the consequences of their use. The authors and publishers have attempted to trace the copyright holders of all material reproduced in this publication and apologize to copyright holders if permission to publish in this form has not been obtained. If any copyright material has not been acknowledged please write and let us know so we may rectify in any future reprint.

Except as permitted under U.S. Copyright Law, no part of this book may be reprinted, reproduced, transmitted, or utilized in any form by any electronic, mechanical, or other means, now known or hereafter invented, including photocopying, microfilming, and recording, or in any information storage or retrieval system, without written permission from the publishers.

For permission to photocopy or use material electronically from this work, access www.copyright. com or contact the Copyright Clearance Center, Inc. (CCC), 222 Rosewood Drive, Danvers, MA 01923, 978-750-8400. For works that are not available on CCC please contact mpkbookspermissions@ tandf.co.uk

Trademark notice: Product or corporate names may be trademarks or registered trademarks, and are used only for identification and explanation without intent to infringe.

Library of Congress Cataloging-in-Publication Data
Names: Prakash, Chander, editor. | Singh, Sunpreet, editor. |
Davim, J. Paulo, editor.
Title: Functional and smart materials / edited by Chander Prakash, Sunpreet
Singh, and J. Paulo Davim.
Other titles: Functional and smart materials (CRC Press)
Description: First edition. | Boca Raton, FL : CRC Press, 2020. |
Series: Manufacturing design and technology | Includes bibliographical
references and index.
Identifiers: LCCN 2020018332 (print) | LCCN 2020018333 (ebook) |
ISBN 9780367275105 (hardback) | ISBN 9780429298035 (ebook)
Subjects: LCSH: Smart materials.
Classification: LCC TA418.9.S62 F85 2020 (print) | LCC TA418.9.S62 (ebook) |
DDC 620.1/12—dc23
LC record available at https://lccn.loc.gov/2020018332
LC ebook record available at https://lccn.loc.gov/2020018333

ISBN: 978-0-367-27510-5 (hbk)
ISBN: 978-0-429-29803-5 (ebk)

Typeset in Times
by codeMantra

Contents

Preface..vii
Editors...ix
Contributors xi

Chapter 1 Electrochromics for Smart Windows: Oxide-Based Thin Films 1

A. Henni, Y. Bouznit, D. Zerrouki, and D. Selloum

Chapter 2 Polymeric Biomaterials in Tissue Engineering.................................. 19

Akhilesh Kumar Maurya and Nidhi Mishra

Chapter 3 Amorphous Semiconductors: Past, Present, and Future 37

Neeraj Mehta

Chapter 4 Promise of Self-lubricating Aluminum-Based
Composite Material 65

Neeraj Kumar Bhoi, Harpreet Singh, and Saurabh Pratap

Chapter 5 Energy Materials and Energy Harvesting... 83

K.S. Smaran, S.G. Patnaik, V. Raman, and N. Matsumi

Chapter 6 Advanced Processing of Superalloys for Aerospace Industries 109

Swadhin Kumar Patel, Biswajit Swain, and Ajit Behera

Chapter 7 Review on Rheological Behavior of Aluminum Alloys in
Semi-Solid State 123

R. Gupta, A. Sharma, and U. Pandel

Chapter 8 Bio-Nanomaterials: An Inevitable Contender in Tissue
Engineering 133

*Pankaj Dipankar, Tara Chand Yadav, Pallavi Saxena, and
Shanid Mohiyuddin*

Chapter 9 Biomaterials 165

Manoj Mittal

v

Chapter 10 Characterisation and Optimisation of TiO_2/CuO Nanocomposite for Effective Dye Degradation from Water under Simulated Solar Irradiation 195

Rohini Singh and Suman Dutta

Chapter 11 Dual Applicability of Hexagonal Pyramid-Shaped Nitrogen-Doped ZnO Composites As an Efficient Photocatalyst 207

Rohini Singh and Suman Dutta

Index 219

Preface

In this book a clear understanding of the *Functional and Smart Materials* is illustrated for various engineering applications by establishing the relationship between their physical properties and the atomic scale structure of materials. Indeed, the development of smart materials will be a vital effort in many fields of science and technology, including automobile, aerospace, microelectronics, computer science, medical treatment, life science, energy, safety engineering, and defense technologies. Further, the materials development in the near-future must be oriented toward the generation of functional materials that can set up new milestones in the different engineering applications such as Biomedical, Aerospace, Automobile, Refineries, Power plants, Marine, Semi-conductors, etc. Current materials research is developing various pathways that lead modern technology toward the smart system.

Each study in this book has focused on the innovative material systems, their development, mechanical and metallurgical characterization, or end-user application(s). Particular emphasis has been placed on the consideration of the predictability, adaptability, and repeatability. The scope of this book is the discovery of new functional materials that will develop smart systems. Moreover, this book presents an intrinsic connection among several dependent systems commonly used in functional materials as well as the evolution of behaviors. Despite sourcing key materials and technologies involved in the development of *Functional and Smart Materials*, this book intends to reveal multidisciplinary scientific rules like geometrical features, physics and chemistry principles, materials science, metrology, process hybridization and optimization, and mechanical and metallurgical responses.

Our aim is to explain the intrinsic connections among different structures of the materials, processes, and operations. The primary goal of this book is to explore new routes for synthesizing functional materials and their use. Ultimately, this book aims to illustrate not only the properties of functional materials but to understand characterization of the obtained materials through advanced measurement techniques.

Editors

Chander Prakash is Associate Professor at the School of Mechanical Engineering, Lovely Professional University, Jalandhar, India. He received a Ph.D. in Mechanical Engineering from Panjab University, Chandigarh, India. His areas of research is biomaterials, rapid prototyping and 3D printing, advanced manufacturing, modeling, simulation, and optimization. He has more than 11 years of teaching experience and 6 years of research experience. He has contributed extensively to titanium- and magnesium-based implant literature with publications appearing in *Surface and Coating Technology, Materials and Manufacturing Processes, Journal of Materials Engineering and Performance, Journal of Mechanical Science and Technology, Nanoscience and Nanotechnology Letters*, and *Proceedings of the Institution of Mechanical Engineers, Part B: Journal of Engineering Manufacture.* He has authored 150 research papers and 30 book chapters. He is also an editor of 15 Books: He is also a guest editor of two journals: Special Issue on "Metrology in Materials and Advanced Manufacturing," *Measurement and Control* (SCI indexed) and Special Issue on "Nano-Composites and Smart Materials: Design, Processing, Manufacturing and Applications" of *Advanced Composites Letters.*

Sunpreet Singh is researcher in NUS Nanoscience & Nanotechnology Initiative (NUSNNI). He received a Ph.D. in Mechanical Engineering from Guru Nanak Dev Engineering College, Ludhiana, India. His area of research is additive manufacturing and application of 3D printing for development of new biomaterials for clinical applications. He has contributed extensively to the subject of additive manufacturing with publications appearing in *Journal of Manufacturing Processes, Composite Part: B, Rapid Prototyping Journal, Journal of Mechanical Science and Technology, Measurement, International Journal of Advance Manufacturing Technology*, and *Journal of Cleaner Production.* He has authored 10 book chapters and monographs. He is working in joint collaboration with Prof. Seeram Ramakrishna, NUS Nanoscience & Nanotechnology Initiative and Prof. Rupinder Singh, Manufacturing Research Lab, GNDEC, Ludhiana. He is also an editor of three books: *Current Trends in Bio-manufacturing*, Springer Series in Advanced Manufacturing, Springer International Publishing AG, Gewerbestrasse 11, 6330 Cham, Switzerland., December 2018; *3D Printing in Biomedical Engineering*, Book series Materials Horizons: From Nature to Nanomaterials, Springer International Publishing AG, Gewerbestrasse 11, 6330 Cham, Switzerland, August 2019; and *Biomaterials in Orthopaedics and Bone Regeneration - Design and Synthesis*, Book series: Materials Horizons: From Nature to Nanomaterials, Springer International Publishing AG, Gewerbestrasse 11, 6330 Cham, Switzerland, March 2019. He is also Guest Editor of three journals: Special Issue on "Functional Materials and Advanced Manufacturing," Facta Universitatis, Series: Mechanical Engineering (Scopus Index), Materials Science Forum (Scopus Index), and Special Issue on "Metrology in Materials and Advanced Manufacturing," *Measurement and Control* (SCI indexed).

J. Paulo Davim received a Ph.D. in Mechanical Engineering in 1997, a M.Sc. degree in Mechanical Engineering (materials and manufacturing processes) in 1991, a Mechanical Engineering degree (five years) in 1986 from the University of Porto (FEUP), the Aggregate title (Full Habilitation) from the University of Coimbra in 2005, and a D.Sc. from London Metropolitan University in 2013. He is Senior Chartered Engineer by the Portuguese Institution of Engineers with an MBA and Specialist title in Engineering and Industrial Management. He is also Eur Ing by FEANI-Brussels and Fellow (FIET) by IET-London. Currently, he is a professor at the Department of Mechanical Engineering of the University of Aveiro, Portugal. He has more than 30 years of teaching and research experience in manufacturing, materials, and mechanical and industrial engineering, with special emphasis in machining and tribology. He has also interest in management, engineering education, and higher education sustainability. He has guided many postdoc, Ph.D. and master's students as well as coordinated and participated in several financed research projects. He has received several scientific awards. He has worked as evaluator of projects for the European Research Council (ERC) and other international research agencies as well as examiner of Ph.D. candidates for many universities in different countries. He is the editor-in-chief of several international journals, a guest editor of journals, books series, and a scientific advisor for many international journals and conferences. Presently, he is an editorial board member of 30 international journals and acts as reviewer for more than 100 prestigious Web of Science journals. He has also published as editor (and co-editor) of more than 100 books and as author (and co-author) of more than 10 books, 80 book chapters, and 400 articles in journals and conferences (more than 250 articles in journals indexed in Web of Science core collection/h-index 49+/7000+ citations, SCOPUS/h-index 56+/10000+ citations, Google Scholar/h-index 70+/16000+).

Contributors

Ajit Behera
Department of Metallurgical and
  Materials Engineering
NIT Rourkela
Rourkela, India

Neeraj Kumar Bhoi
Mechanical Engineering
  Department
Indian Institute of Information
  Technology, Design and
  Manufacturing
Jabalpur, India

Y. Bouznit
Laboratoire de Matériaux :
  Elaborations-Propriétés-Applications
Université de Jijel
Jijel, Algeria

Pankaj Dipankar
Department of Biotechnology
Indian Institute of Technology
Roorkee (U.K.), India

Suman Dutta
Department of Chemical
  Engineering
Indian Institute of Technology (ISM)
Dhanbad, India

R. Gupta
Sigma Carbon Technologies
Jaipur, India

A. Henni
Laboratory Dynamic Interactions and
  Reactivity of Systems
Kasdi Merbah University
Ouargla, Algeria

N. Matsumi
School of Materials Science
Japan Advanced Institute of Science
  and Technology
Nomi, Japan

Akhilesh Kumar Maurya
Chemistry Lab, Department of Applied
  Sciences
Indian Institute of Information
  Technology
Allahabad, India

Neeraj Mehta
Department of Physics, Institute of
  Science
Banaras Hindu University
Varanasi, India

Nidhi Mishra
Chemistry Lab, Department of Applied
  Sciences
Indian Institute of Information
  Technology
Allahabad, India

Manoj Mittal
Department of Mechanical Engineering
IKG Punjab Technical University
Jalandhar, India

Shanid Mohiyuddin
Department of Biotechnology
Indian Institute of Technology
Roorkee (U.K.), India

U. Pandel
Department of Metallurgical and
  Materials Engineering
MNIT Jaipur, Jaipur, India

Swadhin Kumar Patel
Department of Metallurgical and
  Materials Engineering
NIT Rourkela
Rourkela, India

S.G. Patnaik
Laboratory for Analysis and Architecture
  of Systems: LAAS-CNRS
Toulouse, France

Saurabh Pratap
Mechanical Engineering Department
Indian Institute of Information
  Technology, Design and
  Manufacturing
Jabalpur, India

V. Raman
Center for Fuel Cell Technology
International Advanced Research
  Centre for Powder Metallurgy
  and New Materials: ARCI
Chennai, India

Pallavi Saxena
Department of Botany
Mohanlal Sukhadia University
Udaipur, India

D. Selloum
Laboratory Dynamic Interactions
  and Reactivity of Systems
Kasdi Merbah University
Ouargla, Algeria

A. Sharma
Ex. Prof. Department of Metallurgical
  and Materials Engineering
MNIT Jaipur
Jaipur, India

Harpreet Singh
Mechanical Engineering Department
Indian Institute of Information
  Technology, Design and
  Manufacturing
Jabalpur, India

Rohini Singh
Department of Chemical Engineering
Sitarambhai Naranji Patel Institute of
  Technology & Research Centre
Bardoli, India

K.S. Smaran
Department of Chemistry
Sri Sathya Sai Institute of Higher
  Learning
Brindavan Campus, Bengaluru

Biswajit Swain
NIT Rourkela
Rourkela, India

Tara Chand Yadav
Department of Biotechnology
Indian Institute of Technology
Roorkee (U.K.), India

D. Zerrouki
Laboratory Dynamic Interactions
  and Reactivity of Systems
Kasdi Merbah University
Ouargla, Algeria

1 Electrochromics for Smart Windows
Oxide-Based Thin Films

A. Henni
Kasdi Merbah University

Y. Bouznit
Université de Jijel

D. Zerrouki and D. Selloum
Kasdi Merbah University

CONTENTS

1.1 Introduction 1
1.2 The Energy Efficiency of Chromogenic Fenestration 2
1.3 History and Applications 4
1.4 Operating Principles and Materials 4
    1.4.1 Transparent Conductive Electrodes 5
    1.4.2 Electrolyte 6
    1.4.3 Electrochromic Layer 7
1.5 Electrochromic Oxide Films 9
    1.5.1 Anodic Coloured Material: NiO As Reference 9
    1.5.2 Cathodic Coloured Material: WO_3 As Reference 10
1.6 Conclusions 12
References 13

1.1 INTRODUCTION

The first civilizations were built using natural materials: wood, stone, leather, bone, horn, linen or hemp. Recently, an emergence of plastics and composites has been observed in the building, automotive, aeronautic, sport and military sectors. But gradually, researchers and engineers have had to use materials with their own functions. Nanotechnology has modified the landscape of energy generation, energy storage and energy saving devices. Architectural fenestration can extensively benefit from green nanotechnologies. Therefore, our habitat will have to evolve technologically, and this will undoubtedly pass through the use of high-tech windows, called smart, also called "smart windows". These smart windows have the ability to block the heat waves induced by infrared radiation from the sun, and therefore, limit the use of air-conditioning systems while continuing to contribute to the well-being of people working or residing in these buildings.

Amongst them, smart windows are able to control the throughput of visible light and solar radiation into buildings and can impart energy efficiency as well as human comfort by having different transmittance levels depending on dynamic needs. Under the action of a voltage or an electric current, the ECs materials can change their properties. They can be integrated in devices that modulate their transmittance, reflectance, absorptance or emittance. Thus, the electrochromic (EC) phenomenon refers to the formation of a new optical absorption band by the redox process and therefore colour changes in materials. As the redox reaction is reversible, by removing the electric field, the colour of the material will return to the ground state. Electrochromism is known to exist in many types of materials. This chapter deals with devices based on EC metal oxides. There are several promising EC materials such as V_2O_5, [1,2], NiO [3,4] and WO_3 [5,6], which are considered as inorganic or oxide EC material while polyaniline [7] and viologens [8] are common polymer and organic EC materials. The EC phenomenon appears particularly intense by the metal oxides. Of all oxide-based materials, WO_3 has quick response time, high coloration efficiency and long life.

Smart windows are currently used in a growing number of buildings in which some windows are fully colored and others transparent. We note that electrochromism being to the "green" nanotechnologies that are very much in focus today [9].

In this chapter, we will recall a general description of intelligent materials and the different classes. We then describe EC materials in detail, such as their history, interesting properties and some applications of these materials. Also, this chapter discusses oxide-based EC smart windows with emphasis on recent work related to thin films.

1.2   THE ENERGY EFFICIENCY OF CHROMOGENIC FENESTRATION

We start by looking at the world's population and its interrelationship to our common environment. The population has increased from approximately one billion in 1800 to seven billion in 2019. The result of this evolution is that the strains on the global resources are growing steeply and that there is an unsustainable demand on the resources of all kinds: energy, water, minerals, etc. The large use of fossil fuels generates a large amount of CO_2 by causing the global surface temperature by the greenhouse effect. For this, we try to use the technological power for ecological purposes in order to reduce energy consumption. Since most people spend their time inside buildings, a large portion of energy is consumed inside our buildings to ensure our comfort. Indeed, the energy between 30% and 40% [10] is provided to satisfy our needs in air conditioning, heating, ventilation, etc. It is therefore not surprising that the architectural trend converges towards building design with a growing proportion of windows.

However, from a purely energetic point of view, we find the so-called low-emissivity windows (low-e windows): thin enough to maintain optical transparency; in addition, they are thermal insulators and they block a large part of the infrared

Electrochromics for Smart Windows

electromagnetic radiation. However, this static technology is not adequate for the winter if one wants to take advantage of the solar radiation to warm the house.

In this sense, chromogenic materials (materials that modify their optical properties as a result of an external stimulus) are more promising for use as smart windows. These materials include thermochromic materials that are temperature-activated; photochromic materials that are widely used for ophthalmic lenses that opacify due to exposure to ultraviolet light; and EC materials that stain in the face of electrical current.

The technologies based on other thermochromic and photochromic materials do not offer as much flexibility for the user and depend on the external environment and thus offer no control. For example, being sensitive to UV rays, photochromic materials stain in cloudy weather and staining, and discoloration times are affected by ambient temperature.

On the other hand, the use of EC materials is based on the application of a small potential difference necessary to modulate the colouring state of a smart window, and only a reverse voltage is required to return to the degree of transparency. Thus, a power requirement is minimal for their operation. Moreover, we find that some devices using photovoltaic cells [11]. As clearly shown in Figure 1.1, it is found that the EC materials are the most advantageous since they offer a high modulation both in the visible and in the near IR, thus reducing energy costs in lighting, heating and air conditioning. This technology is also suggested for other applications such as the opening roofs of luxury cars or the portholes of some aircraft [12].

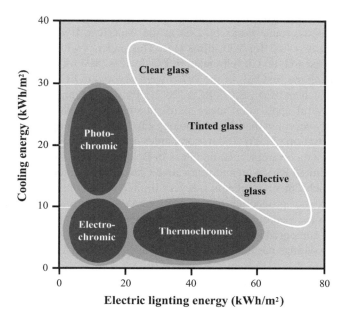

FIGURE 1.1 Electric lighting energy and cooling energy for different types of fenestration[13].

The opacity of an EC window can be controlled by gauging the quantity of charges inserted. The transmittance of the device developed by the Gedimat company varies in the visible range, while making it possible to control the gain in solar heat (the IR radiation associated with heat).

1.3 HISTORY AND APPLICATIONS

The study of EC materials goes back a long way. As early as 1704, Diesbach discovered Prussian blue (ferric ferrocyanide), which is a molecule that will later be widely used as a pigment in several dyes, inks or coatings [14]. Electrochemistry has of course developed since those years and Prussian blue now has applications in electro-catalysis, electro-analysis and obviously in electrochromism.

As early as 1815, Berzelius demonstrated that by heating pure WO_3 in an atmosphere containing hydrogen, it changes from a pale yellow to a dark blue [15]: it thus observes an optical modulation in the visible. Then, Herschel's work on a photographic device involves ferric ferrocyanide molecules that change the colour upon exposure to light. However, in this case it is photochromism rather than electrochromism, but that does not prevent that this method is widely developed. It is interesting to know that the origin of the word "blueprint" comes from the particular coloring associated with this process.

Electrochromism in thin films of metal oxides seems to have been discovered several times through independent work. Compared to the electrochromism of WO_3 in the solid state, it is necessary to wait until the 1953s with the team of Kraus that characterizes the electro-reduction of WO_3 in contact with an acid electrolyte [16]. During the 1960s, some works by Deb at the American Cyanamide Corporation led to analogous results for WO_3 films [17]. This paper is widely seen as the starting point for research and development of EC devices. Several researches in several countries during the first half of the 1970s were devoted to EC devices for information displays such as IBM [18,19], RCA [20,21], Philips [22] and Canon [23]. During the 1980s, several application ideas focus on smart windows technology when it became widely accepted that this technology can be of great importance for energy-efficient fenestration [24,25]. Today, companies developing these devices are numerous, here are some examples: Halio Glass in Belgium, Vario in Canada, Intelligent Glass in the United Kingdom and Innovative Glass in the United States.

The electrochromism market is still very limited for a number of reasons, including the high cost of producing these systems and the various technological barriers that remain to be overcome. Thus, the appearance of EC systems on the economic market is late compared to their discovery in the laboratory. Nevertheless, this type of functional coatings has been commercially available for a few years.

1.4 OPERATING PRINCIPLES AND MATERIALS

An EC system is a multilayer coating whose optical properties are adjustable and reversible when an electric field is applied to its terminals. The principle of these devices is based on the behaviour of an "EC" layer that has the particularity of changing optical properties, from a transparent state to an opaque, absorbing

Electrochromics for Smart Windows

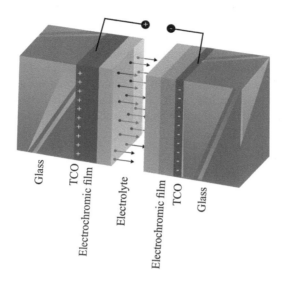

FIGURE 1.2 Principle construction of a foil-based electrochromic device.

state [26]. This transition is due to a colouring/discolouring phenomenon of the layers or multilayer used. Figure 1.2 illustrates a generic EC design [27] with multilayer composed with five layers positioned between two substrates as a laminate. Glass is the most commonly used substrate material, but plastic is an alternative and flexible foil of polyethylene terephthalate (PET) that allows device fabrication by roll-to-roll technology [28].

The main current goals in materials science related to EC are as follows:

1.4.1 Transparent Conductive Electrodes

Generally, the transparent electronic conductive electrodes located on both sides of the system are ITO (SnO_2:In). For several years, this material has been the subject of an absolutely remarkable research work. The ITO electrodes make it possible to establish a potential difference within the system. The thin layers of ITO are usually deposited on a glass substrate. Their absorption in the visible is generally less than 2% and their factor of reflection in the infrared exceeds 90%. They combine low resistivity (~1.10^{-4} Ω/cm) with excellent transmission (over 85%–90%) and good durability. Deposition on flexible substrates can introduce risks of cracking and delamination, inducing a decrease in electrical conductivity. The ITO electrode can be replaced by other transparent semiconductors such as ZnO:Al, ZnO:In, ZnO:Ga or SnO_2:F [29].

Very thin metal films with high transparency, electrical conductivity and higher mechanical pressure resistance than the oxide layers can also be considered as TCOs. Lansaker et al. [30] have proposed a conductive transparent material consisting of a gold layer of very small thickness (<10 nm) inserted between two films of TiO_2. This multilayer had a transmission of 80% and a resistivity of the order of 10^{-4} Ω/cm. Leftheriotis et al. [31] have reported a transparent conductive ZnS/Ag/ZnS material also providing good transmission (>70%) at a wavelength of 550 nm.

Carbon-based coatings are also possible such as nanotube [32,33] and graphene [34,35] and could become interesting in the future for optimisation of EC devices. The features of these materials, such as chemical stability, electrical conductivity and transparency make them a promising electrode alternative for EC device [36].

1.4.2 ELECTROLYTE

The electrolyte is the central part of an EC device and must meet a set of criteria:

- It must be a good ionic conductor to obtain fast kinetics.
- It must not deteriorate in the optimized potential window to colour and discolour the EC device.
- It must be transparent for the application of smart windows.
- It must be chemically inert with electrical insulation.
- It must not degrade, gel or evaporate under the effect of temperature and must be mechanically resistant.
- It must be manufactured at low cost and non-polluting for the environment.
- It must ensure adhesion between the two EC layers.

In an aqueous medium, the electrolytes H_2SO_4 [37] and H_3PO_4 [38] are conventionally used to cyclize WO_3 and the electrolyte KOH [39] to cyclize NiO. For EC devices, there are essentially three types of electrolytes: solid films (metal oxides), polymeric membranes and ionic liquids. Laminated EC devices can be obtained from polymeric membranes and ionic liquids while solid EC devices contain solid films.

The metal oxides Ta_2O_5 [40], ZrO_2 [41], $LiTaO_3$ [42], $LiNbO_3$ [43] are some examples of solid electrolyte meeting many of the criteria. The thin layers are generally porous, and the pre-insertion of H^+ and Li^+ cations is carried out by co-spraying, chemically, electrochemically or by exposure to moisture. For example, Niwa and Takai [44] report an ionic conductivity of 3.10^{-6} S/cm for thin films of Ta_2O_5.

The ionic conductive membranes consist mainly of a polymer (solid or gel) that acts as a host matrix and an ionic salt that provides the ions. The viscosity of the membrane and the concentration of ions present will play a role in the kinetics of the system. These membranes contain monovalent cations such as H^+, Li^+, Na^+ and K^+ [45,46].

The solid polymers are generally composed of a lithium salt, such as $LiClO_4$ or $LiPF_6$, dissolved in a polymer such as poly (methylmethacrylate) (PMMA), polyvinylidene difluroride (PVDF) or poly(ethylene oxide) (PEO). No solvents are used in solid polymers.

The membranes can be gelified and are composed of a large amount of liquid electrolyte in a host polymer. They have a better ionic conductivity but less interesting mechanical properties than solid polymers. The most widely used polymers are polyethylene oxide, poly(acrylonitrile), poly(vinylidene fluoride-hexafluoropropylene) piezoelectric poly(vinylidene fluoride) and poly(methyl methacrylate). The ionic conductivity of solid polymer electrolytes is of the order of 10^{-7} S/cm, whereas it is of the order of 10^{-3} S/cm for the gels [47,48].

The ionic liquids are disseminated in the form of salts whose melting point is less than $100\,^\circ$ C, acting both as a solvent and as a salt. Unlike aqueous electrolytes, their negligible vapour pressures give them very low evaporation.

The non-volatility of the ionic liquid also makes it possible to operate in higher temperature ranges than those used for aqueous solvents. They are non-ignitable, have a high ionic conductivity between 10^{-2} and 10^{-5} S/cm and are stable in a large window of electrochemical potential (about 4 V) [49–52].

There are many combinations of anions and cations to adjust the properties of ionic liquids. The most used cations are imidazolium, ammonium, phosphonium, triazolium, pyrrolidinium and pyridinium. There are also numerous anions such as bis((trifluoromethyl)sulphonyl)imide ($TFSI^-$), perchlorate (ClO_4^-), tetrafluoroborate (BF_4^-), hexafluorophosphate (PF_6^-) and trifluoroacetate ($CF_3CO_2^-$).

1.4.3 ELECTROCHROMIC LAYER

There are mainly three major families of EC materials:

- Conductive polymers.
- Viologen and Prussian blue compounds.
- Metal oxides.

The EC properties of poly-N-methylpyrrole as conductive polymer were first described in 1981 by Diaz et al. for the [53]. Since then, intense work has been done on conductive polymers and these derivatives. Generally, conductive polymers are synthesized organically or by electropolymerisation. The optical properties of these materials can be modified by controlling the doping, thus playing on the band gap between the HOMO energy level and the LUMO energy level [54].

Several research studies have been done on derived materials such as aromatics and aromatic heterocyclic structures. The electrochemical oxidation of aromatic molecules, such as pyrrole, thiophene, aniline, furan, etc., leads to electroactive conductive polymers (oxidation/reduction).

Table 1.1 summarizes some of the EC conductive polymers. Polythiophene [55,56], polypyrrole [57] and polyaniline [58,59] received the most attention with regard to their EC properties. The main advantages of these conductive polymers are their low response times and the possibility of obtaining a large area of colour panel. The main disadvantage is their limited lifetime and instabilities to UV radiation compared to metal oxides.

The second family is the compounds based on viologen and Prussian blue. These compounds have fast response times but do not have a good memory effect.

Prussian blue (iron(III) hexacyanoferrate(II): $Fe_4^{III}Fe^{II}\left[\left(CN\right)_6\right]_3$) accidental discovery in 1704 by Diesbach [60], is the most studied EC metal complex. In reduction, Prussian blue passes successively from yellow to green, to blue then to the transparent state. They can play both the role of cathodic or anodic colouring material. Prussian blue is often deposited on a conductive substrate by electroreduction.

TABLE 1.1

Some Examples of Electrochromic Conductive Polymers [54]

Polymer	Colour of the Polymer (film)	
	Oxidized State (Doped)	Reduced State (Undoped)
Polypyrrole	Brown	Yellow
Polyaniline	Green-blue	Yellow-transparent
Polythiophene	Brown	Green
Polymethylthiophene	Blue	Red
Poly-2,2′-bithiophene	Blue-grey	Red
Poly-3,4-dimethylthiophene	Dark blue	Blue

The most studied viologen is methyl viologen (1,1′-dimethyl-4,4′-bipyridilium). The methyl viologen is mainly known in three states: in neutral form, with a positive charge (radical cation) and with two positive charges (dication). Dication is the most stable form and is colourless, while viologen with a radical cation is the compound with the most intense colour. The viologen is particularly used as EC mirrors by the company Gentex [61].

A number of oxides, especially transition metals, have the ability to change colour with the degree of oxidation of the cation. Table 1.2 shows some examples

TABLE 1.2

The Main Electrochromic Metal Oxides with Anodic and Cathodic Coloration with Their Associated Colours

Type of Colouring	Metal	Oxidized Form	Reduced Form
Cathodic	Bismuth	Bi_2O_3 (Transparent)	$Li_xBi_2O_3$ (Black-brown)
	Molybdenum	MoO_3 (Transparent)	M_xMoO_3 (Dark blue)
	Niobium	Nb_2O_5 (Transparent)	$M_xNb_2O_5$ (blue)
	Titanium	TiO_2 (Transparent)	M_xTiO_2 (Blue-grey)
	Tungsten	WO_3 (Transparent)	M_xWO_3 (Dark blue)
Anodic	Nickel	$NiOOH$ (Brown)	$Ni(OH)_2$ (Transparent)
	Iridium	IrO_2H_2O (Blue-grey)	$Ir(OH)_3$ (Transparent)
	Cobalt	Co_3O_4 (Black-brown)	$M_xCo_3O_4$ (Light yellow)
	Manganese	MnO_2 (Black-brown)	$M_xCo_3O_4$ (Light yellow)
Cathodic and anodic	Vanadium	V_2O_5 (Brown-yellow)	$M_xV_2O_5$ (Light blue)

of inorganic EC metal oxides with cathodic and anodic staining. The oxides of iridium [62], rhodium [63], ruthenium [64], tungsten [65], manganese [66], cobalt [67] or nickel [68] are some examples of metal oxides having EC properties. Vanadium oxide has the particularity of being both materials with cathodic and anodic coloration [69]. The most widely studied EC material is tungsten trioxide (WO_3). It has established itself as the basic EC material for the application of smart windows.

There are many deposition techniques for obtaining these thin-film oxides such as evaporation, sputtering, electrodeposition and sol-gel route [70–77]. The thin layers can be deposited with a small thickness and made in large areas. However, although known to be stable under UV radiation, these oxide deposits may have the disadvantages of being sensitive to moisture and not very resistant to shocks that can lead to cracking.

1.5 ELECTROCHROMIC OXIDE FILMS

A number of oxides, especially transition metals, have the ability to change colour with the degree of oxidation of the cation. There are two principally different kinds of EC oxides: those referred to as "cathodic" and colouring under ion insertion, and the "anodic" ones that colour under ion extraction. Figure 1.3 shows the elements with anodic and cathodic EC properties.

1.5.1 Anodic Coloured Material: NiO As Reference

These materials are typically p-type semiconductors. Iridium oxide was the first compound of this category to be studied, first for its applications in alkaline batteries and then in the field of EC devices [78]. IrO_2 was mainly studied in the 1980s, but because of its price, which has become too high for industrialization and because of its blue colour in the oxidized state, research has focused on nickel oxide (NiO).

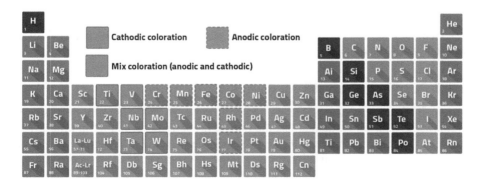

FIGURE 1.3 The periodic table (except the lanthanides and actinides) that shows the transition metals capable of displaying cathodic and anodic electrochromism.

NiO has been studied for more than 30 years. Lampert et al., in 1985, revealed the EC properties of NiO in a basic electrolyte (KOH). The reaction is based on the fact that the transformation of nickel oxide (NiO) into nickel hydroxide ($Ni(OH)_2$) is spontaneous in an alkaline solution. These authors qualified its favourable EC properties for the application of smart windows and associated the colour change of the NiO thin film with the following reaction [79]:

$$Ni(OH)_2 \rightarrow NiOOH + H^+ + e^-$$

The NiO is a brown oxide when oxidized and colourless when reduced. The brown colour correlated with the electro-oxidation of Ni^{2+} to Ni^{3+} and the formation of a nickel oxyhydroxide.

Thin films of NiO can be deposited by various techniques with different morphologies and cycled in various electrolytes in order to obtain the best EC performances. In particular, it has been shown during all these years that porous morphologies were ideal for obtaining the best performances.

Svensson and Granqvist reported the EC properties of a thin layer of hydrated NiO deposited by reactive sputtering. Very satisfactory optical properties over more than 10,000 cycles have been shown with bleached transmission of 75% (at a wavelength of 550 nm), and 15% in the coloured state. The NRA (nuclear reaction analysis) characterization technique allowed the authors to associate the staining process with the extraction of hydrogen ions [80].

In 1987, from the KOH medium cycling of an electrodeposited hydrated NiO layer, Carpenter et al deified the electrochemical reaction as being the same as that which occurs in nickel hydroxide-based batteries [81]. The degradation of the performances after 500 cycles was associated with a transformation of the α-$Ni(OH)_2$ phase in the β-$Ni(OH)_2$ phase and a displacement of the oxidation and reduction peaks towards the positive potentials. Mechanical deterioration was also advanced.

Avendano et al. studied the effect of doping, Mg, Al, Si, V, Zr, Nb, Ag or Ta [82], on the EC properties of sputtered NiO thin films. With the exception of V and Ag, the addition of these elements led to an increase in transparency in the faded state.

In 2008, Xia et al. made porous NiO layers by chemical voice at room temperature, followed by heat treatment [83]. The film annealed at 300°C exhibited a memory effect of several hours, and an optical transmission which varied from 82% between the coloured and faded state. The colour efficiency was 42 mC/cm^2 for a wavelength of 550 nm.

1.5.2 CATHODIC COLOURED MATERIAL: WO_3 AS REFERENCE

Tungsten trioxide has emerged as the basic EC material for the application of smart windows. WO_3 is an oxide that has the optical property of switching in thin layers between the colourless and the blue colour [84]. The reversible change in its optical properties is due to the variation in the degree of oxidation of the tungsten cation. This is due to the insertion (exclusion) of positively charged ions associated with the injection (extraction) of electrons within the material. Insertion and extraction

Electrochromics for Smart Windows

of protons (H+) and electrons (e−) in WO_3 can be described by the highly simplified electrochemical reaction:

$$\left[WO_3 + H^+ + e^-\right]_{bleached} \leftrightarrow \left[HWO_3\right]_{colored}$$

The band gap of WO_3 can vary from 2.4 eV (cubic structure) to 3.12 eV (orthorhombic) leading to a high transparency of WO_3 [85]. The reduction of W^{6+} ions in W^{5+} is associated with the creation of localized states under the conduction band. The electronic transfer of these to the conduction band can then be ensured by the absorption of a low energy photon, corresponding to a red absorption band (≈1,4 eV) and which confers on the material a blue colour in transmission [28] (Figure 1.4).

WO_3 is conventionally cycled in an electrolyte having small cations such as H+ (H_3PO_4 or H_2SO_4) or Li+ ($LiClO_4$-PC or ionic liquid with a lithium salt). The mechanisms involving in these cations are equivalent. WO_3 is most often present in its amorphous form [86] even though the crystallized and/or hydrated forms of WO_3 also exhibit interesting EC properties [87].

The thin layers of WO_3 can be deposited by a wide variety of physical or chemical techniques such as sol-gel [88,89], evaporation [90,91], spray [92,93], electrodeposition [94,95], sputtering [96,97], etc.

Mixed oxides of tungsten may exhibit properties superior to those of pure oxides with additions such as TiO_2 [87, 88, 89], SiO_2 [90], V_2O_5 [91], Ti [92], Mo [93, 94] and Nb [95]. In particular, W-Ti oxide, where the addition of Ti, significantly increases durability during electrochemical cycling. Titanium helps to stabilize a very messy structure. WO_3 has also been associated with polymers such as polyaniline [96] or PEDOT-PSS [97] to obtain multi-colour systems or improve stability. Further optimization of EC oxides can be accomplished in ternary compositions, and a comprehensive study was recently performed on thin films of $W_{1-x-y}Ti_xMo_yO_3$ with $x<0.2$ and $y<0.2$. Figure 1.5 shows chromaticity coordinates for several oxides with Ti contents of ~10 at.% and demonstrates that approximate colour neutrality can be obtained [98].

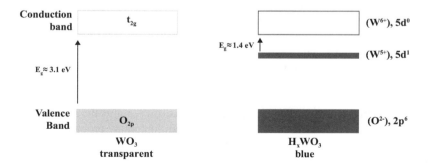

FIGURE 1.4 Simplified diagram explaining the origin of the tungsten oxide.

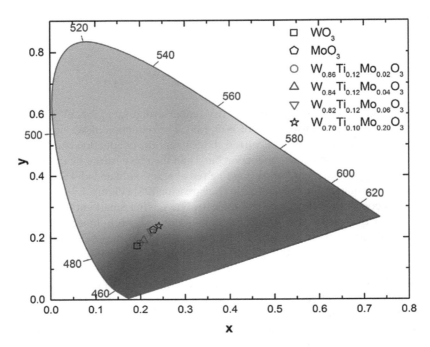

FIGURE 1.5 CIE 1931 chromaticity diagram with (x,y) coordinates for ~300-nm-thick $W_{1-x-y}Ti_xMo_yO_3$ films with the shown compositions. Numbers along the curve denote wavelengths in nm [98].

1.6 CONCLUSIONS

Electrochromism is part of green nanotechnologies that have a major interest in fenestration in energy-efficient buildings. Unlike many other "green" technologies, electrochromism, based on oxides films, is not used the rare earth elements whose availability and cost remains a major problem.

Research on smart windows has grown considerably in the last few years. This type of window is currently manufactured by several companies around the world knowing that low-cost manufacturing and long-term durability are the keys to product success. Lightweight devices in the form of sheets or a glass-adapted roll can be used for ease of placement.

WO_3 is the most studied material in the field of electrochromism and surely the best understood. NiO has been studied for over 30 years and seems to be adequate to be associated with WO_3 because it has good EC performance. However, although it is a simple metal oxide, its electrochemical mechanism remains the subject of debate and controversy, within the EC community. In addition, many problems persist such as durability, temperature resistance and homogeneity of large layers. The improvement of the EC properties of a material undoubtedly depends on the understanding of the mechanism at the origin of its properties. Thus, the coating process can be combined with the continuous lamination of two oxides such as WO_3 and NiO.

REFERENCES

1. I. Mjejri, M. Gaudon, A. Rougier, Mo addition for improved electrochromic properties of V_2O_5 thick films, *Sol. Energy Mater. Sol. Cells.* 198 (2019) 19–25. doi: 10.1016/j.solmat.2019.04.010.
2. M.R. Loi, E.A. Moura, T.M. Westphal, R.D.C. Balboni, A. Gündel, W.H. Flores, M.B. Pereira, M.J.L. Santos, J.F.L. Santos, A. Pawlicka, C.O. Avellaneda, Impact of Zr precursor on the electrochemical properties of V_2O_5 sol-gel films, *J. Electroanal. Chem.* 839 (2019) 67–74. doi: 10.1016/j.jelechem.2019.03.012.
3. K. Zrikem, G. Song, A.A. Aghzzaf, M. Amjoud, D. Mezzane, A. Rougier, UV treatment for enhanced electrochromic properties of spin coated NiO thin films, *Superlattices Microstruct.* 127 (2019) 35–42. doi: 10.1016/j.spmi.2018.03.042.
4. Y.E. Firat, A. Peksoz, Efficiency enhancement of electrochromic performance in NiO thin film via Cu doping for energy-saving potential, *Electrochim. Acta.* (2018). doi: 10.1016/j.electacta.2018.10.166.
5. J. Besnardiere, B. Ma, A. Torres-Pardo, G. Wallez, H. Kabbour, J.M. González-Calbet, H.J. Von Bardeleben, B. Fleury, V. Buissette, C. Sanchez, T. Le Mercier, S. Cassaignon, D. Portehault, Structure and electrochromism of two-dimensional octahedral molecular sieve h'-WO_3, *Nat. Commun.* 10 (2019) 327. doi: 10.1038/s41467-018-07774-x.
6. Q. Zhao, Y. Fang, K. Qiao, W. Wei, Y. Yao, Y. Gao, Printing of WO_3/ITO nanocomposite electrochromic smart windows, *Sol. Energy Mater. Sol. Cells.* 194 (2019) 95–102. doi: 10.1016/j.solmat.2019.02.002.
7. S. Zhang, S. Chen, F. Yang, F. Hu, B. Yan, Y. Gu, H. Jiang, Y. Cao, M. Xiang, High-performance electrochromic device based on novel polyaniline nanofibers wrapped antimony-doped tin oxide/TiO_2 nanorods, *Org. Electron.* 65 (2019) 341–348. doi: 10.1016/j.orgel.2018.11.036.
8. K. Madasamy, D. Velayutham, V. Suryanarayanan, M. Kathiresan, K.-C. Ho, Viologen-based electrochromic materials and devices, *J. Mater. Chem. C.* 7 (2019) 4622–4637. doi: 10.1039/C9TC00416E.
9. G.B. Smith, C.-G.S. Granqvist, *Green Nanotechnology: Solutions for Sustainability and Energy in the Built Environment*, United Kingdom: CRC Press, (2013).
10. P. Huovila, *Buildings and Climate Change: Status, Challenges, and Opportunities*, France: UNEP/Earthprint, (2007).
11. L.-M. Huang, C.-W. Hu, H.-C. Liu, C.-Y. Hsu, C.-H. Chen, K.-C. Ho, Photovoltaic electrochromic device for solar cell module and self-powered smart glass applications, *Sol. Energy Mater. Sol. Cells.* 99 (2012) 154–159. doi: 10.1016/j.solmat.2011.03.036.
12. G. Salek, B. Bellanger, I. Mjejri, M. Gaudon, A. Rougier, Polyol synthesis of Ti-V_2O_5 nanoparticles and their use as electrochromic films, *Inorg. Chem.* (2016). doi:10.1021/acs.inorgchem.6b01662.
13. C.G. Granqvist, P.C. Lansåker, N.R. Mlyuka, G.A. Niklasson, E. Avendaño, Progress in chromogenics: New results for electrochromic and thermochromic materials and devices, *Sol. Energy Mater. Sol. Cells.* 93 (2009) 2032–2039. doi: 10.1016/j.solmat.2009.02.026.
14. P. Monk, R. Mortimer, D. Rosseinsky, *Electrochromism and Electrochromic Devices*, Elsevier, United Kingdom: Cambridge University Press, (2007).
15. R.J. Mortimer, Electrochromic materials, *Annu. Rev. Mater. Res.* 41 (2011) 241–268. doi: 10.1146/annurev-matsci-062910-100344.
16. C.G. Granqvist, *Handbook of Inorganic Electrochromic Materials*, Netherlands: Elsevier, (1995).
17. S.K. Deb, A novel electrophotographic system, *Appl. Opt.* 8 (1969) 192. doi: 10.1364/AO.8.S1.000192.

18. I.F. Chang, Electrochemichromic systems for display applications, *J. Electrochem. Soc.* 122 (1975) 955. doi: 10.1149/1.2134377.
19. I.F. Chang, Electrochromic and electrochemichromic materials and phenomena, in: *Nonemissive Electrooptic Displays*, Boston, MA: Springer US, (1976), pp. 155–196. doi:10.1007/978-1-4613-4289-2_10.
20. B.W. Faughnan, R.S. Crandall, M.A. Lampert, Model for the bleaching of WO_3 electrochromic films by an electric field, *Appl. Phys. Lett.* 27 (1975) 275–277. doi: 10.1063/1.88464.
21. R.S. Crandall, B.W. Faughnan, Dynamics of coloration of amorphous electrochromic films of WO_3 at low voltages, *Appl. Phys. Lett.* 28 (1976) 95–97. doi: 10.1063/1.88653.
22. J. Bruinik, A. Kmetz, F. von Willisen, *Nonemissive Electrooptic Displays*, (1976) 201.
23. Y. Hajimoto, T. Hara, Coloration in a WO_3 film, *Appl. Phys. Lett.* 28 (1976) 228–229. doi: 10.1063/1.88707.
24 C.M. Lampert, Electrochromic materials and devices for energy efficient windows, *Sol. Energy Mater.* 11 (1984) 1–27. doi:10.1016/0165-1633(84)90024-8.
25. J.S.E.M. Svensson, C.G. Granqvist, Electrochromic tungsten oxide films for energy efficient windows, *Sol. Energy Mater.* 11 (1984) 29–34. doi: 10.1016/0165-1633(84)90025-X.
26. C.G. Granqvist, A. Azens, P. Heszler, L.B. Kish, L. Österlund, Nanomaterials for benign indoor environments: Electrochromics for "smart windows", sensors for air quality, and photo-catalysts for air cleaning, *Sol. Energy Mater. Sol. Cells.* 91 (2007) 355–365. doi: 10.1016/j.solmat.2006.10.011.
27. C.G. Granqvist, Electrochromics for smart windows: Oxide-based thin films and devices, *Thin Solid Films.* 564 (2014) 1–38. doi: 10.1016/j.tsf.2014.02.002.
28. C.G. Granqvist, M.A. Arvizu, I.B. Pehlivan, H.Y. Qu, R.T. Wen, G.A. Niklasson, Electrochromic materials and devices for energy efficiency and human comfort in buildings: A critical review, *Electrochim. Acta.* 259 (2018) 1170–1182. doi: 10.1016/j.electacta.2017.11.169.
29. C.G. Granqvist, Transparent conductors as solar energy materials: A panoramic review, *Sol. Energy Mater. Sol. Cells.* 91 (2007) 1529–1598. doi: 10.1016/j.solmat.2007.04.031.
30. P.C. Lansåker, J. Backholm, G.A. Niklasson, C.G. Granqvist, $TiO_2/Au/TiO_2$ multilayer thin films: Novel metal-based transparent conductors for electrochromic devices, *Thin Solid Films.* (2009). doi:10.1016/j.tsf.2009.02.158.
31. G. Leftheriotis, S. Papaefthimiou, P. Yianoulis, Integrated low-emittance-electrochromic devices incorporating ZnS/Ag/ZnS coatings as transparent conductors, *Sol. Energy Mater. Sol. Cells.* (2000). doi: 10.1016/S0927-0248(99)00101-4.
32. F.J. Berger, T.M. Higgins, M. Rother, A. Graf, Y. Zakharko, S. Allard, M. Matthiesen, J.M. Gotthardt, U. Scherf, J. Zaumseil, From broadband to electrochromic notch filters with printed monochiral carbon nanotubes, *ACS Appl. Mater. Interfaces.* 10 (2018) 11135–11142.
33. K. Yanagi, R. Moriya, Y. Yomogida, T. Takenobu, Y. Naitoh, T. Ishida, H. Kataura, K. Matsuda, Y. Maniwa, Electrochromic carbon electrodes: Controllable visible color changes in metallic single-wall carbon nanotubes, *Adv. Mater.* 23 (2011) 2811–2814. doi: 10.1002/adma.201100549.
34. K. Mallikarjuna, H. Kim, Highly transparent conductive reduced graphene oxide/silver nanowires/silver grid electrodes for low-voltage electrochromic smart windows, *ACS Appl. Mater. Interfaces.* 11 (2018) 1969–1978.
35. J. Wu, D. Qiu, H. Zhang, H. Cao, W. Wang, Z. Liu, T. Tian, L. Liang, J. Gao, F. Zhuge, Flexible electrochromic V_2O_5 thin films with ultrahigh coloration efficiency on graphene electrodes, *J. Electrochem. Soc.* 165 (2018) D183–D189.

36. H. Wang, M. Barrett, B. Duane, J. Gu, F. Zenhausern, Materials and processing of polymer-based electrochromic devices, *Mater. Sci. Eng. B Solid-State Mater. Adv. Technol.* 228 (2018) 167–174. doi:10.1016/j.mseb.2017.11.016.
37. J. Wang, E. Khoo, P.S. Lee, J. Ma, Controlled synthesis of WO_3 nanorods and their electrochromic properties in H_2SO_4 electrolyte, *J. Phys. Chem. C.* (2009). doi: 10.1021/jp901650v.
38. H. Yoo, K. Oh, Y.C. Nah, J. Choi, K. Lee, Single-step anodization for the formation of WO_3-doped TiO_2 nanotubes toward enhanced electrochromic performance, *ChemElectroChem.* (2018). doi: 10.1002/celc.201800981.
39 C.Y. Jeong, Y. Abe, M. Kawamura, K.H. Kim, T. Kiba, Electrochromic properties of rhodium oxide thin films prepared by reactive sputtering under an O_2 or H_2O vapor atmosphere, *Sol. Energy Mater. Sol. Cells.* (2019). doi: 10.1016/j.solmat.2019.109976.
40. M. Kitao, H. Akram, K. Urabe, S. Yamada, Properties of solid-state electrochromic cells using Ta_2O_5 as electrolyte, *J. Electron. Mater.* (1992). doi: 10.1007/BF02660405.
41. K.K. Purushothaman, G. Muralidharan, Nanostructured NiO based all solid state electrochromic device, *J. Sol-Gel Sci. Technol.* (2008). doi: 10.1007/s10971-007-1657-0.
42. X. Song, G. Dong, F. Gao, Y. Xiao, Q. Liu, X. Diao, Properties of NiO_x and its influence upon all-thin-film $ITO/NiO_x/LiTaO_3/WO_3/ITO$ electrochromic devices prepared by magnetron sputtering, *Vacuum.* (2015). doi: 10.1016/j.vacuum.2014.09.007.
43. R.B. Goldner, G. Seward, K. Wong, T. Haas, G.H. Foley, R. Chapman, S. Schulz, Completely solid lithiated smart windows, *Sol. Energy Mater.* (1989). doi: 10.1016/0165-1633(89)90020-8.
44. T. Niwa, O. Takai, Optical and electrochemical properties of all-solid-state transmittance-type electrochromic devices, *Thin Solid Films.* (2010). doi: 10.1016/j.tsf.2009.11.062.
45. S.A. Agnihotry, Nidhi, Pradeep, S.S. Sekhon, Li+ conducting gel electrolyte for electrochromic windows, *Solid State Ionics.* (2000). doi: 10.1016/S0167-2738(00)00339-8.
46. O. Bohnke, Gel electrolyte for solid-state electrochromic cell, *J. Electrochem. Soc.* (2006). doi: 10.1149/1.2069512.
47. V.K. Thakur, G. Ding, J. Ma, P.S. Lee, X. Lu, Hybrid materials and polymer electrolytes for electrochromic device applications, *Adv. Mater.* (2012). doi: 10.1002/adma.201200213.
48. S.S. Sekhon, S.A. Agnihotry, Solvent effect on gel electrolytes containing lithium salts, *Solid State Ionics.* (2000). doi: 10.1016/S0167-2738(00)00584-1.
49. R. Marcilla, F. Alcaide, H. Sardon, J.A. Pomposo, C. Pozo-Gonzalo, D. Mecerreyes, Tailor-made polymer electrolytes based upon ionic liquids and their application in all-plastic electrochromic devices, *Electrochem. Commun.* (2006). doi: 10.1016/j.elecom.2006.01.013.
50. H. Moulki, D.H. Park, B.K. Min, H. Kwon, S.J. Hwang, J.H. Choy, T. Toupance, G. Campet, A. Rougier, improved electrochromic performances of NiO based thin films by lithium addition: From single layers to devices, *Electrochim. Acta.* (2012). doi: 10.1016/j.electacta.2012.03.123.
51. W. Lu, A.G. Fadeev, B. Qi, B.R. Mattes, Fabricating conducting polymer electrochromic devices using ionic liquids, *J. Electrochem. Soc.* (2004). doi: 10.1149/1.1640635.
52. A. Branco, L.C. Branco, F. Pina, Electrochromic and magnetic ionic liquids, *Chem. Commun.* (2011). doi: 10.1039/c0cc03892j.
53. A.F. Diaz, J.I. Castillo, J.A. Logan, W.Y. Lee, Electrochemistry of conducting polypyrrole films, *J. Electroanal. Chem.* (1981). doi: 10.1016/S0022-0728(81)80008-3.
54. J. Roncali, Molecular engineering of the band gap of π-conjugated systems: Facing technological applications, *Macromol. Rapid Commun.* (2007). doi: 10.1002/marc.200700345.

55. M.E. Nicho, H. Hu, C. López-Mata, J. Escalante, Synthesis of derivatives of polythiophene and their application in an electrochromic device, *Sol. Energy Mater. Sol. Cells.* (2004). doi: 10.1016/j.solmat.2004.01.009.

56. S. Kula, A. Szlapa-Kula, S. Krompiec, P. Gancarz, M. Filapek, An electrochromic behavior of novel polythiophenes obtained from unsymmetrical monomers- a comprehensive study, *Synth. Met.* (2019). doi: 10.1016/j.synthmet.2018.11.018.

57. C. Dulgerbaki, N.N. Maslakci, A.I. Komur, A.U. Oksuz, Electrochromic strategy for tungsten oxide/polypyrrole hybrid nanofiber materials, *Eur. Polym. J.* 107 (2018) 173–180. doi: 10.1016/j.eurpolymj.2018.07.050.

58. S. Xiong, S. Li, X. Zhang, R. Wang, R. Zhang, X. Wang, B. Wu, M. Gong, J. Chu, Synthesis and performance of highly stable star-shaped polyaniline electrochromic materials with triphenylamine core, *J. Electron. Mater.* 47 (2018) 1167–1175. doi: 10.1007/s11664-017-5901-2.

59. S. Zhang, S. Chen, F. Hu, L. Ding, Y. Gu, B. Yan, F. Yang, M. Jiang, Y. Cao, Patterned flexible electrochromic device based on monodisperse silica/polyaniline core/shell nanospheres, *J. Electrochem. Soc.* 166 (2019) H343. doi: 10.1149/2.1161908jes.

60. M. Ware, Prussian blue: Artists' pigment and chemists' sponge, *J. Chem. Educ.* 85 (2008) 612. doi: 10.1021/ed085p612.

61. R.J. Mortimer, D.R. Rosseinsky, P.M.S. Monk, *Electrochromic Materials and Devices*, Weinheim, Germany: Wiley-VCH Verlag GmbH & Co. KGaA, (2015). doi: 10.1002/9783527679850.

62. G. Beni, L.M. Schiavone, J.L. Shay, W.C. Dautremont-Smith, B.S. Schneider, Electrocatalytic oxygen evolution on reactively sputtered electrochromic iridium oxide films [7], *Nature.* (1979). doi: 10.1038/282281a0.

63. S. Gottesfeld, The anodic rhodium oxide film: A two-color electrochromic system, *J. Electrochem. Soc.* (1980). doi: 10.1149/1.2129654.

64. S.H. Lee, P. Liu, H.M. Cheong, C. Edwin Trecy, S.K. Deb, Electrochromism of amorphous ruthenium oxide thin films, *Solid State Ionics*, 165 (2003) 217–221. doi: 10.1016/j.ssi.2003.08.035.

65. H.Y. Qu, E.A. Rojas-González, C.G. Granqvist, G.A. Niklasson, Potentiostatically pretreated electrochromic tungsten oxide films with enhanced durability: Electrochemical processes at interfaces of indium–tin oxide, *Thin Solid Films.* 682 (2019) 163–168. doi: 10.1016/j.tsf.2019.02.027.

66. S.I. Córdoba De Torresi, A. Gorenstein, Electrochromic behaviour of manganese dioxide electrodes in slightly alkaline solutions, *Electrochim. Acta.* 37 (1992) 2015–2019. doi: 10.1016/0013-4686(92)87117-I.

67. A. El Bachiri, L. Soussi, O. Karzazi, A. Louardi, A. Rmili, H. Erguig, B. El Idrissi, Electrochromic and photoluminescence properties of cobalt oxide thin films prepared by spray pyrolysis, *Spectrosc. Lett.* 52 (2019) 66–73. doi: 10.1080/00387010.2018.1556221.

68. M.Z. Sialvi, R.J. Mortimer, G.D. Wilcox, A.M. Teridi, T.S. Varley, K.G.U. Wijayantha, C.A. Kirk, Electrochromic and colorimetric properties of nickel(II) oxide thin films prepared by aerosol-assisted chemical vapor deposition, *ACS Appl. Mater. Interfaces.* 5 (2013) 5675–5682. doi: 10.1021/am401025v.

69. J. Chu, Z. Kong, D. Lu, W. Zhang, X. Wang, Y. Yu, S. Li, X. Wang, S. Xiong, J. Ma, Hydrothermal synthesis of vanadium oxide nanorods and their electrochromic performance, *Mater. Lett.* 166 (2016) 179–182. doi: 10.1016/j.matlet.2015.12.067.

70. A. Henni, A. Merrouche, L. Telli, A. Karar, Studies on the structural, morphological, optical and electrical properties of Al-doped ZnO nanorods prepared by electrochemical deposition, *J. Electroanal. Chem.* 763 (2016) 149–154. doi: 10.1016/j.jelechem.2015.12.037.

71. D. Selloum, A. Henni, A. Karar, A. Tabchouche, N. Harfouche, O. Bacha, S. Tingry, F. Rosei, Effects of Fe concentration on properties of ZnO nanostructures and their application to photocurrent generation, *Solid State Sci.* 92 (2019) 76–80. doi: 10.1016/j.solidstatesciences.2019.03.006.
72. A. Mahroug, B. Mari, M. Mollar, I. Boudjadar, L. Guerbous, A. Henni, N. Selmi, Studies on structural, surface morphological, optical, luminescence and Uv photodetection properties of sol–gel Mg-doped ZnO thin films, *Surf. Rev. Lett.* 26 (2018) 1850167. doi: 10.1142/S0218625X18501676.
73. Y. Bouznit, A. Henni, Characterization of Sb doped SnO_2 films prepared by spray technique and their application to photocurrent generation, *Mater. Chem. Phys.* 233 (2019) 242–248. doi: 10.1016/j.matchemphys.2019.05.072.
74. A. Henni, A. Merrouche, L. Telli, A. Karar, Optical, structural, and photoelectrochemical properties of nanostructured In-doped ZnO via electrodepositing method, *J. Solid State Electrochem.* (2016). doi: 10.1007/s10008-016-3190-y.
75. A. Henni, A. Merrouche, L. Telli, A. Azizi, R. Nechache, Effect of potential on the early stages of nucleation and properties of the electrochemically synthesized ZnO nanorods, *Mater. Sci. Semicond. Process.* 31 (2015) 380–385. doi: 10.1016/j.mssp.2014.12.011.
76. A. Henni, A. Merrouche, L. Telli, A. Karar, F.I. Ezema, H. Haffar, Optical, structural, and photoelectrochemical properties of nanostructured In-doped ZnO via electrodepositing method, *J. Solid State Electrochem.* 20 (2016) 2135–2142. doi: 10.1007/s10008-016-3190-y.
77. G.J. Fang, Z.L. Liu, Y. Wang, Y.H. Liu, K.L. Yao, Synthesis and structural, electrochromic characterization of pulsed laser deposited vanadium oxide thin films, *J. Vac. Sci. Technol. A Vacuum, Surf. Film.* (2001). doi: 10.1116/1.1359533.
78. S. Gottesfeld, J.D.E. McIntyre, G. Beni, J.L. Shay, Electrochromism in anodic iridium oxide films, *Appl. Phys. Lett.* (1978). doi: 10.1063/1.90277.
79. G.F. Cai, C.D. Gu, J. Zhang, P.C. Liu, X.L. Wang, Y.H. You, J.P. Tu, Ultra fast electrochromic switching of nanostructured NiO films electrodeposited from choline chloride-based ionic liquid, *Electrochim. Acta.* 87 (2013) 341–347. doi: 10.1016/j.electacta.2012.09.047.
80. J.S.E.M. Svensson, C.G. Granqvist, Electrochromic hydrated nickel oxide coatings for energy efficient windows: Optical properties and coloration mechanism, *Appl. Phys. Lett.* (1986). doi: 10.1063/1.97281.
81. M.K. Carpenter, R.S. Conell, D.A. Corrigan, The electrochromic properties of hydrous nickel oxide, *Sol. Energy Mater.* (1987). doi: 10.1016/0165-1633(87)90082-7.
82. E. Avendaño, A. Azens, G.A. Niklasson, C.G. Granqvist, Electrochromism in nickel oxide films containing Mg, Al, Si, V, Zr, Nb, Ag, or Ta, *Sol. Energy Mater. Sol. Cells,* (2004). doi: 10.1016/j.solmat.2003.11.032.
83. X.H. Xia, J.P. Tu, J. Zhang, X.L. Wang, W.K. Zhang, H. Huang, Morphology effect on the electrochromic and electrochemical performances of NiO thin films, *Electrochim. Acta.* 53 (2008) 5721–5724. doi: 10.1016/j.electacta.2008.03.047.
84. S.K. Deb, Optical and photoelectric properties and colour centres in thin films of tungsten oxide, *Philos. Mag.* (1973). doi: 10.1080/14786437308227562.
85. A. Rougier, F. Portemer, A. Quédé, M. El Marssi, Characterization of pulsed laser deposited WO_3 thin films for electrochromic devices, *Appl. Surf. Sci.* (1999). doi: 10.1016/S0169-4332(99)00335-9.
86. J.G. Zhang, D.K. Benson, C.E. Tracy, S.K. Deb, A.W. Czanderna, C. Bechinger, Chromic mechanism in amorphous WO_3 films, *J. Electrochem. Soc.* (1997). doi: 10.1149/1.1837737.

87. S.H. Lee, R. Deshpande, P.A. Parilla, K.M. Jones, B. To, A.H. Mahan, A.C. Dillon, Crystalline WO_3 nanoparticles for highly improved electrochromic applications, *Adv. Mater.* (2006). doi: 10.1002/adma.200501953.

88. Y. Ren, Y. Gao, G. Zhao, Facile single-step fabrications of electrochromic WO_3 micro-patterned films using the novel photosensitive sol-gel method, *Ceram. Int.* (2015). doi: 10.1016/j.ceramint.2014.08.084.

89. B. Wen-Cheun Au, K.-Y. Chan, D. Knipp, Effect of film thickness on electrochromic performance of sol-gel deposited tungsten oxide (WO_3), *Opt. Mater. (Amst).* 94 (2019) 387–392. doi: 10.1016/j.optmat.2019.05.051.

90. M.M. El-Nahass, M.M. Saadeldin, H.A.M. Ali, M. Zaghllol, Electrochromic properties of amorphous and crystalline WO_3 thin films prepared by thermal evaporation technique, *Mater. Sci. Semicond. Process.* (2015). doi: 10.1016/j.mssp.2014.02.051.

91. D. Evecan, E. Zayim, Highly uniform electrochromic tungsten oxide thin films deposited by e-beam evaporation for energy saving systems, *Curr. Appl. Phys.* (2019). doi: 10.1016/j.cap.2018.12.006.

92. A. Kumar, C.S. Prajapati, P.P. Sahay, Modification in the microstructural and electrochromic properties of spray-pyrolysed WO_3 thin films upon Mo doping, *J. Sol-Gel Sci. Technol.* (2019). doi: 10.1007/s10971-019-04960-1.

93. R. Mukherjee, P.P. Sahay, Improved electrochromic performance in sprayed WO_3 thin films upon Sb doping, *J. Alloys Compd.* (2016). doi: 10.1016/j.jallcom.2015.11.138.

94. S. Poongodi, P.S. Kumar, D. Mangalaraj, N. Ponpandian, P. Meena, Y. Masuda, C. Lee, Electrodeposition of WO_3 nanostructured thin films for electrochromic and H_2S gas sensor applications, *J. Alloys Compd.* (2017). doi: 10.1016/j.jallcom.2017.05.122.

95. S. Xie, Z. Bi, Y. Chen, X. He, X. Guo, X. Gao, X. Li, Electrodeposited Mo-doped WO_3 film with large optical modulation and high areal capacitance toward electrochromic energy-storage applications, *Appl. Surf. Sci.* (2018). doi: 10.1016/j.apsusc.2018.08.045.

96. A. Subrahmanyam, A. Karuppasamy, Optical and electrochromic properties of oxygen sputtered tungsten oxide (WO_3) thin films, *Sol. Energy Mater. Sol. Cells.* (2007). doi: 10.1016/j.solmat.2006.09.005.

97. V.R. Buch, A.K. Chawla, S.K. Rawal, Review on electrochromic property for WO_3 thin films using different deposition techniques, *Mater. Today Proc.* (2016). doi: 10.1016/j.matpr.2016.04.025.

98. M.A. Arvizu, G.A. Niklasson, C.G. Granqvist, Electrochromic $W_{1-x-y}Ti_xMoyO_3$ thin films made by sputter deposition: Large optical modulation, good cycling durability, and approximate color neutrality, *Chem. Mater.* (2017). doi: 10.1021/acs.chemmater.6b05198.

2 Polymeric Biomaterials in Tissue Engineering

Akhilesh Kumar Maurya and Nidhi Mishra
Indian Institute of Information Technology

CONTENTS

2.1 Introduction 19
2.2 Polymeric Biomaterials 21
    2.2.1 Natural Polymeric Biomaterials 21
        2.2.1.1 Proteins 21
        2.2.1.2 Polysaccharides... 23
    2.2.2 Synthetic Polymeric Biomaterials 25
        2.2.2.1 Polyacrylates 26
        2.2.2.2 Polyesters 26
        2.2.2.3 Poly(Ortho-Esters) 26
        2.2.2.4 Poly(Alkylene Oxalates) 28
        2.2.2.5 Poly(Glycolic Acid).. 28
        2.2.2.6 Poly(Lactic Acid) 28
        2.2.2.7 Other Synthetic Biomaterials... 28
2.3 Application of Polymeric Biomaterials in Tissue Engineering 28
    2.3.1 Bone Regeneration... 29
    2.3.2 Skin Regeneration... 30
    2.3.3 Cardiovascular Tissue Engineering... 31
2.4 Conclusion 31
References... 31

2.1 INTRODUCTION

Mainly tissue injuries originate from physical accident or degenerative diseases. Tissue loss has been provocation in the current scenario because of the shortage of therapeutic resources [1]. In the past decade, the removal of the damaged body part was the very common practice, which resulted in a number of restrictions of the affected individual and in a significant reduction in quality of life [1,2]. The increase in life expectancy has resulted from the finding of new methods to replace/regenerate damaged or degenerated tissues in the body. Thus, replacement and/or regeneration of the damaged body part have become the new target. Thousands of patients have been suffering from tissue loss or organ failure such as liver, kidney, teeth, bone, muscles, etc., and to treat this kind of loss, we need to develop an alternative method; therefore, concept of tissue engineering arisen [3]. Biomaterials are natural or synthetic materials that interact with the biological system. Biomaterials are used in medical applications to augment or replace natural function. Biomaterials used in tissue engineering may be broadly classified as (i) synthetic polymers and (ii) naturally occurring polymers. Synthetic polymers include hydrophobic materials such as α hydroxyl acids, polyanhydrides, etc., and naturally occurring polymer includes complex sugars (hyaluronan) and inorganics (hydroxyapatite) [4]. Polymers have contributed toward the improvement in quality of life in medical science to millions of people worldwide [5,6]. Features such as biocompatibility, cell adhesion, shaping into different forms, cell differentiation support and cell proliferation, ability to mimic the microenvironment of the natural tissues, allowing diffusion of nutrients and residuals have made both natural and synthetic polymers, very attractive materials in tissues regeneration applications [7]. Besides their mechanical characteristics, properties such as surface morphology, as well as the ability of polymers to be degraded into low molecular mass species, have made them materials of choice. Polymeric biomaterials play a significant role in biomedical applications [8,9].

In this chapter, we focus on synthetic and natural polymeric biomaterials and their application in tissue engineering. (See Figure 2.1)

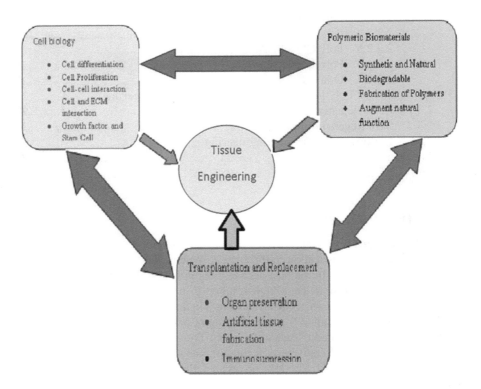

FIGURE 2.1 The relation between tissue engineering, medical science, and research.

Polymeric Biomaterials in Tissue Engineering

2.2 POLYMERIC BIOMATERIALS

Polymeric biomaterials are substances that have been engineered to interact with biological systems for a medical/therapeutic purpose, e.g., treat, repair, or replace a tissue function of the body. Polymeric biomaterials are different from biological material/biopolymers, such as bone, that is produced by a body system. A polymeric biomaterial that is biocompatible or suitable for one application may not be biocompatible in another [10]. The major assets of the biopolymers compared to metal or ceramic materials are (i) easily manufactured and produce variety of shapes such as films, fibers, spheres, gels, sheets, hydrogels, membranes, capsules and 3D-structures (scaffolds) and (ii) secondary processabilities are very easy, acceptable cost and availability with required mechanical and physical properties. The required properties of polymeric biomaterials for therapeutic use are biocompatibility, sterilizability, convenient physical and mechanical properties, and manufacturability. The major types of polymeric biomaterials for biomedical applications are obtained from natural or synthetic organic sources [11].

Polymeric biomaterials are the largest class of biomaterials, and it can be divided on the basis of source (i) natural and (ii) synthetic.

2.2.1 NATURAL POLYMERIC BIOMATERIALS

Nature has developed few basic substances, mainly polymers and minerals for the biological system with remarkable functional properties [12]. Something that is present in and/or produced by nature is natural polymers and is not man-made [13]. Natural polymers can come from anywhere; generally, they originate from microbes, plants/animals, or abiotic sources [14]. Naturally derived polymers can be considered as the first biodegradable and biocompatible biomaterials used clinically [15]. Natural polymeric biomaterials owing to the biologically active properties have better interactions with the cells of body, which permit them to increase the cell's performance in biological system. Naturally derived polymers can be classified as proteins (fibrinogen, silk, gelatin, elastin, keratin, actin, collagen, and myosin), polysugars (cellulose, amylose, dextran, chitin, and glycosaminoglycans), and polynucleotides (DNA and RNA) [16].

2.2.1.1 Proteins

2.2.1.1.1 Collagen

Collagen is the major constituent of the tissues and is the most abundant protein in the animal, about 25% of the total mass of proteins. It is present in the fibrous/connective tissues and extracellular matrix (ECM), and it furnishes shape and structural integrity and strength to connective tissues including skin, bone, tendons, cartilage, blood vessels, and ligaments [17,18]. Twenty-nine types of collagen have been identified [19]. Collagen consists of protein with three polypeptide chain, each polypeptide contains a repeat Gly-X-Y amino acid sequence, where X and Y are generally proline and hydroxyproline, respectively. These polypeptide chains are aligned in a parallel fashion and left-handed coiled polyproline II-type (PPII-type) helix, and triple helix structure is formed by enfolding of it around each other and interstrand hydrogen stabilize helical structure of collagen polymer [20]. Collagen can be divided into several groups according to the structure they form: Type I, II, III, V, and XI and it has quaternary structures. Types IX, XII, XIV, XIX and XXI have fibril-associated collagens with interrupted triple helices (FACIT). Types XV and XVIII have multiple triple helix domains with interruptions (multiplexin). Collagen in the skin, vessels, and tendons is Type I, Type II is the main collagen in cartilage tissues, and collagen in the wall of blood vessels is Type III. Conversely, collagen in the basement membrane which separates epithelial tissue of mesoderm is Type IV and collagen Type V cell surfaces, hair, and placenta [17,21].

Polypeptide chains in a collagen triple helix are made up of thousands of amino acid residues. Glycine (Gly) is every first residue, resulting in a repeating unit of Gly-X-Y, where X and Y can be any amino acid. Repetition of Gly ensures a close packaging tropocollagen triple helix with three strands of polypeptide [22,23]. (See Figure 2.2)

Only a few collagens are used in the fabrication of collagen-based polymeric biomaterials within various types of collagen. Tropocollagen triple helices fabricate into fibrils, which aggregate to form fibers and contain the very commonly used forms of collagen for wound dressing/healing and other tissue engineering purposes. The most abundant type of collagen in animals is Type I and is commonly used for the medicinal purpose [19,24].

2.2.1.1.2 Silk

Silk is a natural polymeric biomaterial that is made up of protein fiber. Silk protein fibers are mainly composed of fibroin and are synthesized by certain larvae of insect to form cocoons [25]. Mulberry silkworm, i.e., *Bombyx mori* and spiders of *Nephila* genus (*N. clavipes, N. madagascarensis*) provide the best-known silk by cocoons of the larvae. Fibrous silks are synthesized by the epithelial cells of these living insects [19]. Silk synthesized by the silkworm consists of two main proteins, sericin, and fibroin. Fibroin is made up of the amino acids Gly-Ser-Gly-Ala-Gly-Ala and forms β-pleated structure due to a 59-mer amino acid repeat with variation in some sequence [26]. Hydrogen bond forms between polypeptide chains, and side chains form hydrogen bond network above and below the plane [27]. Fibroin has a very interesting application in the medical field, and serisin is used to obtain polymeric biomaterials and used in tissue engineering [28]. (See Figure 2.3)

2.2.1.1.3 Fibronectin

Fibronectin (FN) is a multifunctional, ECM glycoprotein and present on the surface of cell membrane, in plasma of blood and other body fluids, molecular weight of

$$\left[\text{Gly-X-Y-Gly-X-Y-Gly-X-Y-Gly-X-Y}\right]_n$$

X and Y May be any
amino acid

FIGURE 2.2 Basic sequence of amino acids in polypeptide chain of collagen.

$$\left[\text{Gly-Ser-Gly-Ala-Gly-Ala}\right]_n$$

FIGURE 2.3 Sequence of amino acids present in silk fibroin.

glycoprotein in the ECM is high (~250 kDa), that can interact with other ECM molecules such as collagen, fibrinogen, and glycosaminoglycans [29]. FN is produced by a wide variety of epithelial and mesenchymal cells: hepatocytes, fibroblasts, chondrocytes, myoblasts, macrophages, and intestinal epithelial cells [30]. FN is considered as a main glycoprotein of the ECM, and it participates in the formation of both structure and shape of cells and functional properties performed by the living tissues in the body [31]. A wide variety of cellular interactions is performed by FN with ECM and plays important roles in cell migration, adhesion, growth, proliferation, and differentiation [31,32].

2.2.1.2 Polysaccharides

Biomaterials of polysaccharides are widely distributed in nature. These materials possess unique structures and characteristic features and have vast importance in different fields. For example, cellulose, chitin, agars, alginate, etc.

2.2.1.2.1 Chitin and Chitosan

Chitin is a natural polysaccharide made up of a single monomer unit of N-acetyl-D-glucosamine (NAG) [33]. Chitin occurs in nature as ordered macrofibrils, and it is a major component of exoskeletons of arthropods and provides structural support, such as the crustaceans (crabs and shrimps), insects, the radulae of mollusks, cephalopod beaks, and the scales of fish and amphibians as well as the cell walls of fungi [34,35]. The structure of chitin $(C_8H_{13}O_5N)_n$ is very similar to that of cellulose, but monomer units are chitin is 2-acetamido-2-deoxy-β-D-glucose (NAG), which are attached to each other via $\beta(1\rightarrow4)$ linkages. Deacetylated form of chitin is chitosan (that can have varying degrees of deacetylation), and it is soluble in acidic solutions [36,37]. For biomedical applications, chitin is converted to its deacetylated derivative, chitosan either by enzymatic preparations or chemical hydrolysis [35,38–41]. Different forms of chitin are synthesized by the action of the enzyme for medical application. (See Figure 2.4)

2.2.1.2.2 Alginate

Alginate is a naturally occurring anionic polymer of saccharide distributed widely in the class of Phaeophyceae, including *Laminaria hyperborea*, *L. digitata*, *L. japonica*, *Ascophyllum nodosum*, and *Macrocystis pyrifera* [42]. α-d-mannuronic acid and β-l-guluronic acid forms the structure of alginate that is derived from seaweed (Phaeophyceae) [43]. Ionic cross-linking in the presence of number of divalent cations, e.g., Ca^{2+}, Mg^{2+}, forms hydrogel of alginate [44]. Alginate has also been covalently cross-linked and oxidized in an attempt to optimize the physical properties of alginate hydrogel. ECMs of living tissues are structurally very similar, and this similarity allows it to be used in medical application such as wound healing, delivery of small chemical drugs and proteins and cell transplantation. Physiological moist

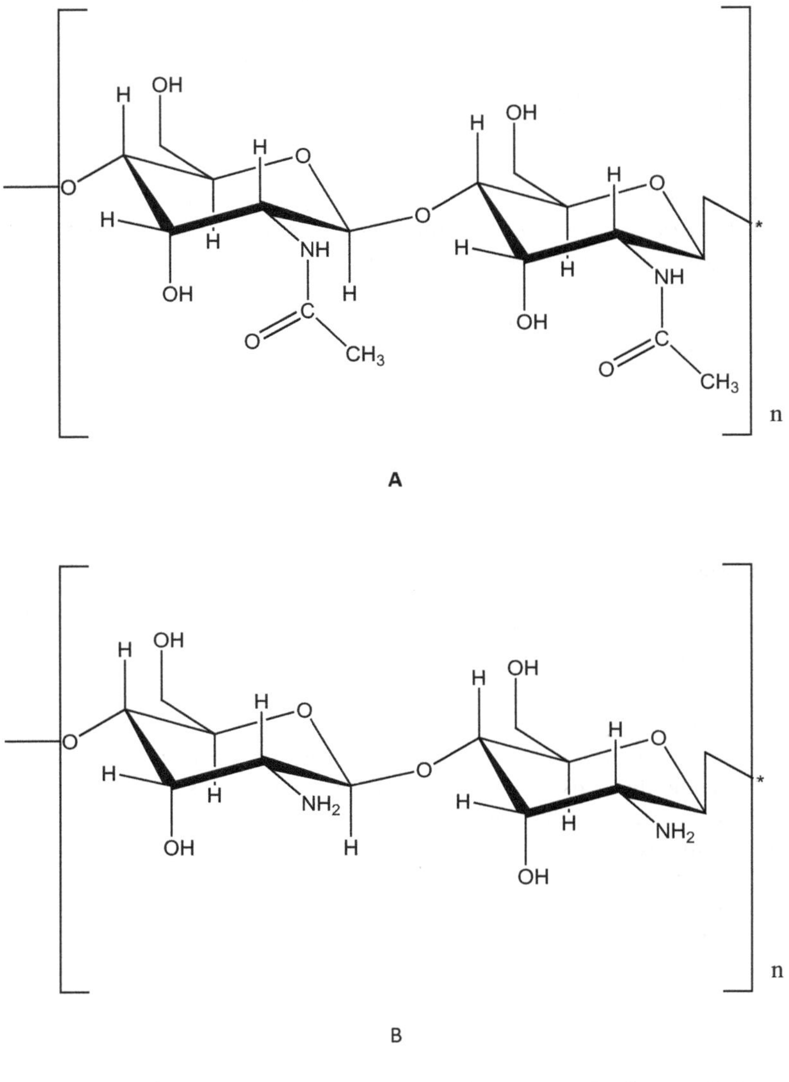

A

B

FIGURE 2.4 (a) Chitin (N-acetyl glucosamine and (b) chitosan (D-glucosamine)

in the microenvironment is maintained by the hydrogel of alginate at site wound and acts as a barrier for microorganisms infection at the site of wound and promotes wound healing [44,45]. (See Figure 2.5)

2.2.1.2.3 Hyaluronan

Hyaluronan is an elastoviscous fluid containing hylan polymers that are derivatives of hyaluronan, a naturally occurring complex sugar [46]. Hyaluronan also called hyaluronic acid (HA) is an anionic, nonsulfated glycosaminoglycan of the ECM distributed widely throughout connective, epithelial, and neural tissues, constituted by a

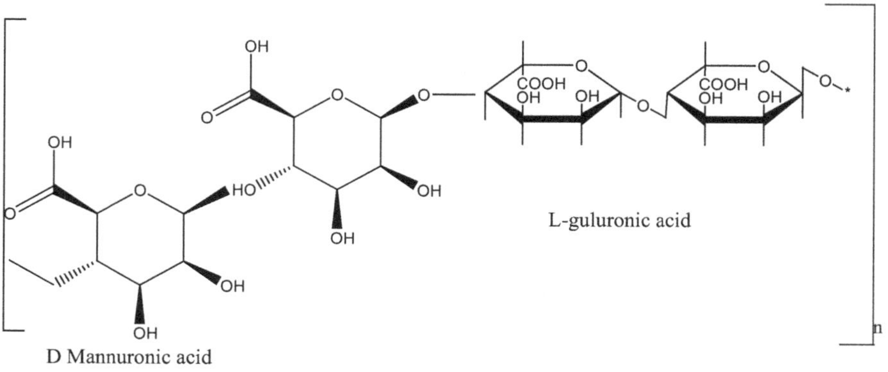

Polymeric Biomaterials in Tissue Engineering

FIGURE 2.5 Alginate with their structural units (β-D-mannuronic acid and α-L-guluronic acid).

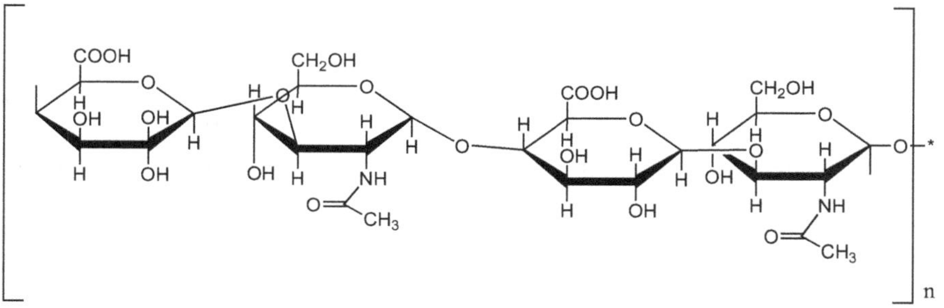

FIGURE 2.6 Hyaluronan consisting of glucouronic acid and N-acetylglucosamine disaccharide.

variable number of repeating glucuronic acid and N-acetylglucosamine disaccharide subunits [47]. Hyaluronan plays an important role in cell migration, cell proliferation, matrix assembly, and tissue development and may be involved in the progression of some malignant tumors [48]. (See Figure 2.6)

2.2.1.2.4 Other Natural Polymeric Biomaterials

In nature, many polymers exist with their specific medical application, such as starch, inulin, guar gum, xanthan gum, psyllium, etc., are polymers of sugars and wheat gum, soy protein, etc. are polymeric proteins and nucleic acids, e.g., DNA and RNA are polymers of nucleotides.

2.2.2 SYNTHETIC POLYMERIC BIOMATERIALS

Synthetic or artificial polymers are human-made polymers. A wide variety of man-made polymers are available with different side chains and main chain. Synthetic polymeric biomaterials have the potential to improve the quality of life; therefore, it has great importance in the medical field [49]. Biochemists are now capable of synthesizing number of synthetic biopolymers. Synthetic polymers offer several advantages over their natural counterparts, including improved chemical resistance, mechanical durability and tunability of their properties.

Mainly four types of synthetic polymeric biomaterials are used in biomaterials that are [50]:

1. Biodegradable materials, which have hydrophilic and nonhydrolytic properties such as poly-ethylene-glycol (PEG), poly-vinyl-alcohol (PVA), and poly-acrylamide (PAAm).
2. Hydrophobic polymers, such as poly(n-butyl acrylate), as well as biomaterials that are hydrophobically and hydrolytically susceptible such as poly-(α-esters).
3. Amphiphilic block polymers such as (PEG-b-PPO-b-PEG).
4. Thermally sensitive polymers such as poly(N-isopropylacrylamide) (pNIPAAM) have also been widely used as biomaterials.

2.2.2.1 Polyacrylates

Polyacrylates are containing a vinyl group and a carboxylic acid ester terminus or a nitril [51]. Polyacrylates are made up of monomer unit of acrylate, which have vinyl groups with ester. With increasing chain length of the alcohol moiety in the polymer of acrylates decreases their hardness and solvent resistance. Copolymer of acrylate with methacrylate, methylmethacrylate or styrene has increase stiffness. Methyl acrylate monomers are used for increasing hydrophilicity, and styrene or 2-ethylhexyl acrylate can be used for increasing hydrophobicity. Styrenated acrylics are tight binders, hydrophobic, which are relatively cheap [52]. Many polymers have been synthesized using acrylate as monomer with other monomers such as poly(2-hydroxyethyl methacrylate) PHEM, poly(N-isopropylacrylamide) (PNIPAAm), poly(tert-butyl acrylate), poly(ethylhexyl acrylate), etc., which are used as biomaterials in medical field. (See Figure 2.7)

2.2.2.2 Polyesters

Polyesters are a polymer that has ester linkage in their key chain. Polyester includes naturally occurring polyesters such as the cutin of plant cuticles, as well as synthesized polyesters such as polybutyrate. Natural polyesters and some synthetic polymer of polyesters are biodegradable. Nowadays, polyesters are developed to generate an enormous interest because of their applicability in the biomedical field [53], such as poly(glycerol-sebacate) (PGS), poly-ε-caprolactone (PCL), poly(lactic-co-glycolic acid) (PLGA), polyhydroxyalkanoates (PHAs), etc.

2.2.2.3 Poly(Ortho-Esters)

Poly(ortho-esters) (POEs) are the polymers of ortho-ester groups along with the main chain. An ortho-ester is a functional group containing three alkoxy groups linked with single carbon atom [54]. POEs are derived by the transesterification of ortho-esters and by polyaddition between a diol and a diketene acetal [55]. (See Figure 2.8) POEs are divided into four types [56]:

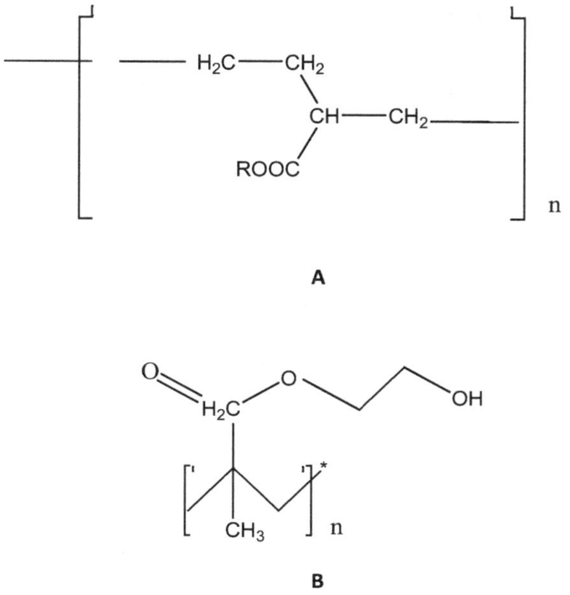

Polymeric Biomaterials in Tissue Engineering

FIGURE 2.7 (a) General structure of polyacrylate and (b) poly (2-hydroxy ethyl methyl acrylate) is an example of polyacrylate.

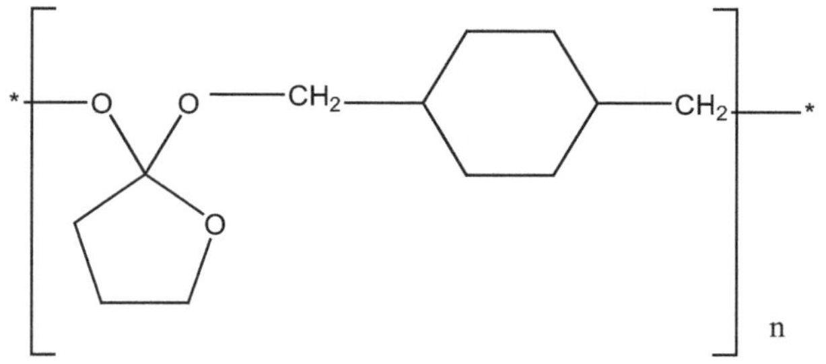

FIGURE 2.8 Monomeric unit of poly(ortho-ester).

- Poly-ortho-ester (POE I).
- Poly-ortho-ester (POE II).
- Poly-ortho-ester (POE III).
- Poly-ortho-ester (POE IV).

POE I are the polymers, which is synthesized from diol and 2, 2′-dimethoxyfuran. POE II polymers are synthesized from a diketene acetal and a diol, POE III polymers are synthesized from triol and a 1,1,1-triethoxy compound and in synthesis of POE IV materials mono/dilactide included in the backbone with POE II reaction [56]. Hydrophobicity of POEs is more than polyesters and polyanhydrides due to lack of polar carbonyl functional group. Surface degradation of POEs occurs very frequently hence it has numerous applications in drug delivery [57].

2.2.2.4 Poly(Alkylene Oxalates)

Polymer of alkylene oxalates are prepared by reaction of dialkyl oxalate with an alkylene diol. It has good tensile properties and a high level of softness and flexibility. Nonabsorbable polymers such as polypropylene, polyethylene, or polytetrafluoroethylene interact with alkylene oxalates and form surgical prosthesis and used in tissue repairing [58]. Many such polymers have been synthesized such as poly (methylene oxalate), poly(ethylene oxalate), poly(butylene oxalate), etc.

2.2.2.5 Poly(Glycolic Acid)

Polyglycolic acid (PGA) is the simplest linear thermoplastic biodegradable polymer and the member of aliphatic polyesters. PGA was the first synthetic absorbable and biodegradable suture introduced in the early 1970s [59]. Polyglycolide or PGA is a simple and linear aliphatic polyester, which can be reverted to carbon dioxide and water under human physiological environmental condition [60,61]. It can be prepared from glycolic acid by means of polycondensation.

2.2.2.6 Poly(Lactic Acid)

Carothers in 1932 discovered polylactic acid (PLA) or poly-lactide. PLA is a thermally stable aliphatic polyester synthesized from natural resources, such as starch of corn, roots of tapioca, starch, sugarcane, etc. [62].Two main monomers are used to synthesize PLA: (i) lactic acid and (ii) cyclic di-ester lactide. Ring-opening polymerization of lactide with various metal catalysts in aqueous is the most common pathway for PLA synthesis [63]. Production of PLA is based on agricultural (crop growing), biological (fermentation) and chemical (polymerization) sciences and technologies. It is classified as generally recognized as safe (GRAS) by the US Food and Drug Administration and is safe for all food packaging applications [64].

2.2.2.7 Other Synthetic Biomaterials

Number of synthetic polymers of biomaterial used in tissue engineering. A small list of synthetic polymers used in tissue engineering is given in Table 2.1.

2.3 APPLICATION OF POLYMERIC BIOMATERIALS IN TISSUE ENGINEERING

Natural biopolymers are ideal for the fabrication of scaffolds for tissue engineering applications to regenerate tissue or organs of the body. Natural polymeric biomaterials are very useful in tissue engineering applications, some of which are bioelectrodes, dental implantation, orthopedic applications, adhesive and sealant, intraocular lens implants and skin substitutes, among others [65].

Compared to naturally derived biopolymers, synthetic biopolymers can be fabricated under more controllable conditions. So, their biological, physical and chemical properties, e.g., mechanical strength, microstructure and degradation rate, are more predictable. Functional groups of the polymers decide the properties of synthesized polymers [66].

TABLE 2.1
Some Synthetic Polymers and Their Monomer Unit

S.L.	Name(s)	Formula	Monomer
1	Polyethylene high density (HDPE)	$-(CH_2-CH_2)_n-$	Ethylene $CH_2=CH_2$
2	Polystyrene (PS)	$-[CH_2-CH(C_6H_5)]_n-$	Styrene $CH_2=CHC_6H_5$
3	Poly(vinyl acetate) (PVAc)	$-(CH_2-CHOCOCH_3)_n-$	Vinyl acetate $CH_2=CHOCOCH_3$
4	Polychloroprene (cis + trans) (neoprene)	$-[CH_2-CH=CCl-CH_2]_n-$	Chloroprene $CH_2=CH-CCl=CH_2$
5	Poly(vinyl chloride) (PVC)	$-(CH_2-CHCl)_n-$	Vinyl chloride $CH_2=CHCl$
6	Polyethylene low density (LDPE)	$-(CH_2-CH_2)_n-$	Ethylene $CH_2=CH_2$
7	Poly(vinylidene chloride) (Saran A)	$-(CH_2-CCl_2)_n-$	Vinylidene chloride $CH_2=CCl_2$
8	cis-Polyisoprene natural rubber	$-[CH_2-CH=C(CH_3)-CH_2]_n-$	Isoprene $CH_2=CH-C(CH_3)=CH_2$
9	Polyacrylonitrile (PAN, orlon, acrilan)	$-(CH_2-CHCN)_n-$	Acrylonitrile $CH_2=CHCN$
10	Polytetrafluoroethylene (PTFE, Teflon)	$-(CF_2-CF_2)_n-$	Tetrafluoroethylene $CF_2=CF_2$

2.3.1 BONE REGENERATION

Tissue engineering for bone regeneration requires artificial scaffolds with physico-chemical, biological and structural properties that can mimic the natural environment of ECM, which gives a proper environmental condition for bone regeneration by cell division and differentiation. Ideal polymeric scaffolds should not show the immuno-suppressant reactions in complex and sensitive biological system; polymers should be in a controlled degradability with non-toxic byproducts and can be easily excreted from body system through metabolic reaction. [66]

Two major naturally synthesized biodegradable polymers, i.e., polysaccharides and proteins, are used in bone tissue engineering, [67]. Collagens are a major fibrous protein in the natural bone tissue. Type I collagen provides the most suitable environment for osteogenesis [68]. Tight interaction between the surface of calcium phosphate and Ca-binding moieties of collagen forms composite which enhance the mechanical strength and rigidity [69].

An anionic glycosaminoglycan is HA with better viscosity, elasticity, and water solubility that has been mainly used as a carrier In recombinant bone morphogenetic protein-2 (rhBMP-2) polymers of HA [70].

Gelatin is cost effective and has greater water solubility than collagen [67]. Gelatin-based composites including gelatin/tricalcium phosphate and methacryloyl/

hydroxyapatite hydrogel scaffolds have better osteoconductive performance due to improved mechanical strength [71].

Tissue engineering for bone chitosans is one of the best bioactive biomaterials [30], and it is used for the management of bone diseases such as osteoporosis and arthritis [72]. Combination of chitosan and inorganics such as calcium phosphates fabricates stronger scaffolds with improved, biocompatible, porous structure, biodegradable, suitable for cell growth, and osteoconductive properties for bone tissue engineering [73]. Composite of chitosan and hydroxyapatite shows identical physical and chemical properties with the organic portion as well as the inorganic portion of natural bone [74].

Biodegradable polymers of urethanes (PUs) are synthetic polymer which is used in bone tissue engineering. Compared to other biodegradable synthetic polymers, PUs possesses better mechanical properties. Molecular weight, chemical composition, and structure determine the properties of PUs. [75].

Hydroxylation of poly-vinyl acetate produces PVA. PVA have the higher degree of hydroxylation property leading to lower water solubility and crystallinity [76].

Generally, acidic poly(amino acids), such as poly(aspartic acid) and γ-poly(glutamic acid) (γ-PGA), and basic poly(amino acids), such as polylysine and polyarginine, are polypeptides which are used in bone or bone tissue regeneration [77].

Polyanhydrides can easily synthesize with desirable characteristics [78]. Biocompatibility and degradability into non-toxic byproducts of polyanhydrides are favorable properties for tissue engineering. They have been developed for bone tissue engineering scaffolds [79].

Poly-ethylene glycol (PEG) is the most studied biopolymer for tissue engineering research. Adhesion of progenitor cells of endothelia dynamically supported by the peptide-grafted PEG hydrogels. [80]. PEG hydrogels are not degradable but can be synthesized as degradable PEG in the form of diblock, triblock, and multiblock copolymers of PL(G)A/PEG [81,82] and can be used in bone regeneration.

2.3.2 Skin Regeneration

Collagens have been widely used as vehicles for transportation of collagen-based implantation of skin or cultured skin cells for skin replacement [83]. For skin replacement, Type I collagen is suitable due to their mechanical strength and biocompatibility [84]. Artificial skin was developed by the modified sponge with the combination of fibrillar collagen with gelatin, dehydrothermal crosslinks are used as stabilizing agent [85].

Fibrin-based biopolymers for skin regeneration have been tested *in vitro* and *in vivo* with PGA [86]. For the proliferation and spreading of human Mesenchymal Stem Cells (hMSCs), fibrin scaffolds are good matrix, which enhance their suitability for wound healing and skin substitution [87].

Polyurethane (PU) is used in dressings because of its ability to provide good barrier for microorganism and permeability to oxygen [88].

PEG has the control over structural and compositional properties; however, it lacks cell-interactive character due to its bio-inert nature [89]. Poly(ethylene glycol-terephthalate) and poly (butylenes terephthalate) are used as dermal substitute [90].

Polymeric Biomaterials in Tissue Engineering

Different amounts of β-d-mannuronic acid and α-l-guluronic acid residues form linear polysaccharides known as alginates. Alginate is a biodegradable polymer and has controllable porosity; therefore, it can enhance cell survival and growth [91].

2.3.3 CARDIOVASCULAR TISSUE ENGINEERING

Gels of polymeric collagens have been developed as bioengineered cardiovascular tissues, such as heart valves, blood vessels, and ligaments [92]. Aggregation of platelets cannot be activated by monomeric collagen, while polymeric collagen having a regular arrangement with a length of around 1 µm does activate aggregation [93].

Fibrin gel is biodegradable and nontoxic biopolymer, so it is suitable for cardiac tissue engineering. Birla et al. have reported that neonatal cardiomyocytes populated with fabricated hollow fibrin gel tubes and implanted into the femoral artery of adult rats. After three weeks of implantation, dense capillary network was formed with mature cardiac tissue [94].

Vascular endothelial growth factor (VEGF), fibroblast growth factor (FGF), and a number of cytokines from which FN bind and perform biological function. First time, McManus et al. cultured neonatal rat cardiac fibroblasts onto fibrinogen electrospun scaffolds [95].

Another polymer that is used for the construction of heart valves is PU-based electrospun materials. For example, electrospun primary cardiac ventricular cells were cultured [96]. Cells grown on aligned PU had significantly more mature state, suggesting that the physical organization of microfibers in PU scaffolds impacts both multicellular architecture and cardiac cell phenotype *in vitro*.

Polyethyleneoxide (PEO) has been extensively evaluated for its thrombin-resistance property. Copolymer of PEO/polypropileneoxide was shown to have high resistance against adhesion of platelets, and mechanical properties are similar to a natural vessel. [97,98].

2.4 CONCLUSION

Polymeric biomaterials are considered as materials that are widely used in medical and pharmaceutical practices. Tissue engineering has been very easy due to the presence of natural as well as synthetic polymers with similar physical, chemical, and biological properties of the body system. Many polymers behave like ECM of tissues which are essential for the cell proliferation, differentiation, and growth of tissue and organ. Tissue engineering is an emerging field of medical science and has broad applications. Besides polymeric biomaterials, a wide range of compounds and/ or biomaterials are present in nature and synthesized which have a variety of uses in tissue engineering.

REFERENCES

1. Santos Jr, A.R. *Bioresorbable Polymers for Tissue Engineering*. InTech Published online 01, March 2010.
2. Hench, L.L. Biomaterials: A forecast for the future. *Biomaterials*, Vol. 19, pp. 1419–1423, 1998.

3. Hutmache, D.W., Goh, J.C.H., Teoh, S.H. An introduction to biodegradable materials for tissue engineering applications. *Biomaterials for Tissue Engineering*, Vol. 30, March 2001.
4. Kohane, D.S., Langer, R. Polymeric biometerials in tissue engineering. *International Pediatric Research Foundation*, Vol. 63, No. 5, pp. 487–489, 2008.
5. Binyamin, G., Shafi, B.M., Mery, C.M. Biomaterials: A primer for surgeons. *Seminars in Pediatric Surgery*, Vol. 15, No. 4, pp. 276–283, November 2006.
6. Dobrzanski, L.A. Significance of materials science for the future development of societies. *Journal of Materials Processing Technology*, Vol. 175, pp. 133–148, 2006.
7. Porjazosa ujundzisi, A., Chamovska, D. *Biodegradable Polymers Suitable for Tissue Engineering and Drug Delivery System*. 2017, E-ISSN 2466-2585.
8. Mano, J.F., souse, R.A., Boesel, L.F., Neves, N.M., Reis, R.L., Biointert, biodegradable and injectable polymeric matrix composites for hard tissue replacement: State of the art and recent developments. *Composites Science and Technology*, Vol. 64, No. 6, pp. 789–817, 2004.
9. Rezaie, H.R., Bakhtiari, L., Öchsner, A. *Biomaterials and Their Applications*. New York: Springer, 2015.
10. Schmalz, G., Arenholdt-Bindslev, D. Chapter 1: Basic aspects. *Biocompatibility of Dental Materials*. Berlin: Springer-Verlag. pp. 1–12, 2008, ISBN 9783540777823. Archived from the original on 9 December 2017. Retrieved 29 February 2016.
11. Dos Santos, V., Brandalise, R.N., Savaris, M. Biomaterials characteristics and properties. *Engineering of Biomaterials*, pp. 8–11. Retrieved 12 August 2017.
12. Fratzl, P. Biomimetic materials research: What can we really learn from nature's structural materials? *Journal of the Royal Society Interface*, Vol. 4, No. 15, pp. 637–642, 2007.
13. Schoental, R. Toxicology of natural produce. *Food and Cosmetics Toxicology*, Vol. 3, No. 4, pp. 609–620, 1965.
14. Nakanishi, K. An historical perspective of natural products chemistry. *Comprehensive Natural Products Chemistry*, Vol. 8, pp. 21–48, 1999.
15. Nair, L.S., Laurencin, C.T. Biodegradable polymers as biomaterials. *Progress in Polymer Science*, Vol. 32, No. 8–9, pp. 762–798, 2007.
16. Yannas, I.V. Classes of materials used in medicine: natural materials in *Biomaterials Science, An Introduction to Materials in Medicine*, Ratner, B.D., Hoffman, A.S., Schoen, F.J., and Lemons, J., Eds., pp. 127–136, Elsevier Academic Press, San Diego, CA, 2004.
17. Loureiro Das Santos, L.A. *Natural Polymeric Biomaterials: Processing and Properties*, 2017.
18. Amoabediny, G. et al. *Biomaterials Science & Engineering*, Pignatello R. Ed., ISBN:978-953-307-609-6, 2011.
19. Chattopadhyay, S., Raines, R.T. Collagen based biomaterials for wound healing. *Biopolymers*, Vol. 101, pp. 821–833, 2014.
20. Okuyama, K., Hongo, C., Fukushima, R., Wu, G., Narita, H., Noguchi, K., Tanaka, Y., Nishino, N. *Biopolymers*, Vol. 76, pp. 367–377, 2004.
21. Gelse, K., Poschl, E., Aigner, T. Collagens—structure, function, and biosynthesis. *Advanced Drug Delivery Reviews*, Vol. 55, pp. 1531–1546, 2003.
22. Anson, M.L., Edsall, J.T. *Advance in Protein Chemistry*. New York: Academic Press INC. Vol. IX, 1954.
23. Preston, R.D. *Advances in Botanical Research*. New York: Academic Press INC, Vol. 2, 1965.
24. Timpl, R. *Extracellular Matrix Biochemistry*. Piez, K.A., Reddi, A.H., Eds., New York: Elsevier, pp. 159–190, 1984.
25. Walker, A.A., Weisman, S., Church, J.S., Merritt, D.J., Mudie, S.T., Sutherland, T.D. Silk from crickets: A new twist on spinning. *PLoS One*, Vol. 7, No. 2, p. e30408, 2012.

Polymeric Biomaterials in Tissue Engineering

26. Menachem, L., Ed., *Handbook of Fiber Chemistry*, 3rd ed., CRC Press, ISBN 0-8247-2565-4, 2006.
27. Muffly, T.M.[1], Tizzano, A.P., Walters, M.D. The history and evolution of sutures in pelvic surgery. *Journal of the Royal Society of Medicine*, Vol. 104, No. 3, pp. 107–12, March 2011.
28. MacIntosh, A.C., Kearns, V.R., Crawford, A., Hatton, P.V. Skeletal tissue engineering using silk biomaterials. *Journal of Tissue Engineering and Regenerative Medicine*, Vol. 2, No. 2, pp. 71–80, 2008.
29. Amoabediny, G. et al. *Biomaterials Science & Engineering*, Pignatello R. Ed., ISBN:978-953-307-609-6, 2011.
30. Sitterley, G. Fibronectin. *BioFiles*, Vol. 3, No. 8, 2008.
31. Clark, R.A.F. *The Molecular and Cellular Biology of Wound Repair*, 2nd ed., New York: Springer Science and Business Media, 1988.
32. Mrsny, R.J, Daugherty, A. *Protein and Peptides Pharmacokinetic, Pharmacodianemic and Metabolic Outcome*, Vol. 202, CRC Press, 2010.
33. Kozlowski, R.M. *Hand Book of Natural Fibres, Vol. 2: Processing and Applications*. New Delhi: Woodhead Publishing Ltd., 2012.
34. Tang, W.J., Fernandez, J.G., Sohn, J.J., Amemiya, C.T. Chitin is endogenously produced in vertebrates. *Current Biology*, Vol. 25, pp. 897–900, 2015.
35. Azuma, K., Izumi, R., Osaki, T., et al. Chitin, chitosan, and its derivatives for wound healing: Old and new materials. *Journal of Functional Biomaterials*, Vol. 6, No. 1, pp. 104–42, 2015.
36. Park, B.K., Kim, M.M. Applications of chitin and its derivatives in biological medicine. *International Journal of Molecular Sciences*, Vol. 11, No. 12, pp. 5152–64, 2010.
37. Shahidi, F., Arachchi, J.K.V., Jeon, Y.J. Food applications of chitin and chitosan. *Trends in Food Science Technology*, Vol. 10, pp. 37–51, 1999.
38. Kafetzopoulos, D., Martinou, A., Bouriotis, V. Bioconversion of chitin to chitosan: purification and characterization of chitin deacetylase from Mucor rouxii. *Proceedings of the National Academy of Sciences of the United States of America*, Vol. 90, No. 7, pp. 2564–8, 1993.
39. Bhagavan, N.V. *Simple Carbohydrates, Medical Biochemistry*, 4th Ed., pp. 133–151, 2002.
40. Aiba, S. Preparation of N-acetylchitooligosaccharides by hydrolysis of chitosan with chitinase followed by N-acetylation. *Carbohydrate Research*, Vol. 265, No. 2, pp. 323–8, 1994.
41. Younes, I., Rinaudo, M. Chitin and chitosan preparation from marine sources. Structure, properties and applications. *Marine Drugs*, Vol. 13, No. 3, pp. 1133–74, 2015.
42. Smidsrod, O., Skjak-Bræk, G. Alginate as immobilization matrix for cells. *Trends in Biotechnology*, Vol. 8, pp. 71–78, 1990.
43. Jiang, T., Singh, B., Choi, Y.-J., Akaike, T., Cho, C.-S. Liver tissue engineering using functional marine biomaterials, *Functional Marine Biomaterials Properties and Application*, Woodhead Publishing Series in Biomaterials, pp. 91–106, 2015.
44. Lee K.Y., Mooney D.J. Alginate: properties and biomedical applications. *Progress in Polymer Science*, Vol. 37, No. 1, pp. 106–126, 2012.
45. Burdick, J.A., Stevens, M.M. *Biomedical Hydrogels. Biomaterials, Artificial Organs and Tissue Engineering*, Woodhead Publishing Series in Biomaterials, pp. 107–115, 2005.
46. Lyn Weiss, M.D., FAAPMR, FAANEM. Chapter 2- medications and injection supplies. *Easy Injections*, pp. 8–14, 2007.
47. Waugh, D.J., McClatchey, A., Montgomery, N., McFarlane, S. Adhesion and penetration: Two sides of CD44 signal transduction cascades in the context of cancer cell metastasis. *Hyaluronan in Cancer Biology*, pp. 109–125, 2009.

48. Numata, K., Kaplan, D.L. Biologically derived scaffolds. *Advanced Wound Repair Therapies*, Woodhead Publishing Series in Biomaterials, pp. 524–551, 2011.
49. Finkenstadt, V.L. Natural polysaccharides as electroactive polymers. *Applied Microbiology and Biotechnology*, Vol. 67, pp. 735–745, 2005.
50. Baldwin, A.D., & Kiick, K.L. Polysaccharide-modified synthetic polymeric biomaterials. Peptide Science: Original Research on Biomolecules, 94(1), 128–140, (2010).
51. Wnek, G., Bowlin, G.L., *Encyclopedia of Biomaterials and Biomedical Engineering*, Informa Health Care, 2004.
52. Baldwin, A.D., Kiick, K.L. Polysaccharide-modified synthetic polymeric biomaterials. *Biopolymers*, Vol. 94, No. 1, pp. 128–140, 2010.
53. Ohara, T., Sato, T., Shimizu, N., Prescher, G., Schwind, H., Weiberg, O., Marten, K., Greim, H. Acrylic acid and derivatives. *Ullmann's Encyclopedia of Industrial Chemistry*. Weinheim: Wiley-VCH, 2002.
54. Chapman, R.A. Chemical bonding. *Handbook of Nonwovens*, Woodhead Publishing Series in Textiles, pp. 330–367, 2007.
55. Mogosanu, D.-E., Giol, E., Vandenhaute, M. et al. Polyester biomaterials for regenerative medicine. *Frontiers in Biomaterials*, Vol. 1, pp. 155–197, 2014.
56. Jeong, B. Injectable biodegradable materials. *Injectable Biomaterials. Science and Applications*, Woodhead Publishing Series in Biomaterials, pp. 323–337, 2011.
57. Casalini, T., Perale, G. Types of bioresorbable polymers for medical applications. *Durability and Reliability of Medical Polymers*, Woodhead Publishing Series in Biomaterials, pp. 3–29, 2012.
58. Murthy, N., Wilson, S., Sy, J.C., Aqida, S.N. Biodegradation of polymers. *Reference Module in Materials Science and Materials Engineering*, 2017.
59. Shalaby, S.W., Damiolkowski, D.D. *Synthetic Absorbable Surgical Devices of Poly(Alkylene Oxalates)* U.S. patent US-4140678-A, February, 20, 1979.
60. Chu, C.C. Materials for absorbable and nonabsorbable surgical sutures. *Biotextiles as Medical Implants*, Woodhead Publishing Series in Textiles, pp. 275–334, 2013.
61. Gilding, D.K., Reed, A.M. Biodegradable polymers for use in surgery - polyglycolic/ poly (lactic acid) homo- and copolymers: 1. *Polymer*, Vol. 20, No. 12, pp, 1459–1464.
62. Ayyoob, M., Lee, D.H., Kim, J.H., Nam, S.W., Kim, Y.J. Synthesis of poly(glycolic acids) via solution polycondensation and investigation of their thermal degradation behaviors. *Fibers and Polymers*, Vol. 18, No. 3, pp, 407–415, 2017.
63. Jamshidian, M., Tehrany, E.A., Imran, M., Jacquot, M., Desobry, S. Poly-lactic acid: Production, applications, nanocomposites, and release studies. *Reviews in Food Science & Food Safety*, August 2010.
64. Södergård, A., Stolt, M. Industrial production of high molecular weight poly(lactic acid). In Auras, R.A., Lim, L.-T., Selke, S.E.M., Tsuji, H. *Poly(Lactic Acid): Synthesis, Structures, Properties, Processing, and Applications*, 2010.
65. Conn, R.E., Kolstad, J.J., Borzelleca, J.F., Dixler, D.S., Filer, L.J., LaDu, B.N., Pariza, M.W. Safety assessment of polylactide (PLA) for use as a food-contact polymer. *Food and Chemical Toxicology,* Vol. 33, pp. 273–83, 1995.
66. Ige, O.O., Umoru, L.E., Aribo, S. Natural products: A minefield of biomaterials. *International Scholarly Research Network ISRN Materials Science*, 2012, Article ID 98306.
67. Shi, C., Yuan, Z., Han, F., Zhu, C., and Li, B., Polymeric biomaterials for bone regeneration. Published: 25 November 2016.
68. Pina, S., Oliveira, J.M., Reis, R.L. Natural-based nanocomposites for bone tissue engineering and regenerative medicine: A review. *Advance Matter*, Vol. 27, pp. 1143–69, 2015.
69. Mizuno, M., Shindo, M., Kobayashi, D., et al. Osteogenesis by bone marrow stromal cells maintained on type I collagen matrix gels in vivo. *Bone*, Vol. 20, pp. 101–7, 1997.
70. Zhang, Y., Reddy, V.J., Wong, S.Y., et al. Enhanced biomineralization in osteoblasts on a novel electrospun biocomposite nanofibrous substrate of hydroxyapatite/collagen/ chitosan. *Tissue Engineering Part A*, Vol. 16, pp. 1949–60, 2010.

Polymeric Biomaterials in Tissue Engineering

71. Hunt, D.R., Jovanovic, S.A., Wikesjo, U.M., Wozney, J.M., Bernard, G.W. Hyaluronan supports recombinant human bone morphogenetic protein-2 induced bone reconstruction of advanced alveolar ridge defects in dogs. A pilot study. *Journal of Clinical Periodontology*, Vol. 72, pp. 651–658, 2001.
72. Visser, J., Gawlitta, D., Benders, K.E., et al. Endochondral bone formation in gelatin methacrylamide hydrogel with embedded cartilage-derived matrix particles. *Biomaterials*, Vol. 37, pp. 174–82, 2015.
73. Porporatto, C., Canali, M.M., Bianco, I.D., Correa, S.G. The biocompatible polysaccharide chitosan enhances the oral tolerance to type II collagen. *Clinical and Experimental Immunology*, Vol. 155, pp. 79–87, 2009.
74. Dhivya, S., Saravanan, S., Sastry, T.P., et al. Nanohydroxyapatite-reinforced chitosan composite hydrogel for bone tissue repair in vitro and in vivo. *Journal of Nanobiotechnology,* Vol. 13, p. 40, 2015.
75. Adav, A.V., Bhise, B. Chitosan: A potential biomaterial effective against typhoid. *Current Science*, Vol. 87, No. 9, pp. 1176–78, 2004.
76. Huang, M.N., Wang, Y.L., Luo, Y.F. Biodegradable and bioactive porous polyurethanes scaffolds for bone tissue engineering. *Journal of Biomedical Science and Engineering*, Vol. 2 pp. 36–40, 2009.
77. Baker M.I., Walsh S.P., Schwartz Z., et al. A review of polyvinyl alcohol and its uses in cartilage and orthopedic applications. *Journal of Biomedical Materials Research Part B: Applied Biomaterials*, Vol. 100, pp. 1451–7, 2012.
78. Hsieh, C.Y., Tsai, S.P., Wang, D.M., et al. Preparation of gamma-PGA/chitosan composite tissue engineering matrices. *Biomaterials*, Vol. 26, pp. 5617–23, 2005.
79. Torres, M.P., Vogel, B.M., Narasimhan, B., Mallapragada, S.K. Synthesis and characterization of novel polyanhydrides with tailored erosion mechanisms. *Journal of Biomedical Materials Research Part A*, Vol. 76, pp. 102–110, 2006.
80. Jain, J.P., Chitkara, D., Kumar, N. Polyanhydrides as localized drug delivery carrier: An update. *Expert Opinion on Drug Delivery*, Vol. 5, pp. 889–907, 2008.
81. Mann, B.K., West, J.L. Cell adhesion peptides alter smooth muscle cell adhesion, proliferation, migration, and matrix protein synthesis on modified surfaces and in polymer scaffolds. *Journal of Biomedical Materials Research Part A*, Vol. 60, pp. 86–93, 2002.
82. Abebe, D.G., Fujiwara, T. Controlled thermoresponsive hydrogels by stereocomplexed PLA-PEG-PLA prepared via hybrid micelles of pre-mixed copolymers with different PEG lengths. *Biomacromolecules*, Vol. 13, pp. 1828–1836, 2012.
83. Buwalda, S.J., Calucci, L., Forte, C., Dijkstra, P.J., Feijen, J. Stereocomplexed 8-armed poly(ethylene glycol)-poly(lactide) star block copolymer hydrogels: Gelation mechanism, mechanical properties and degradation behavior. *Polymer*, Vol. 53, pp. 2809–2817, 2012.
84. Bell, E., Sher, S., Hull, B., Merril, C., Rosen, S., Chamson, A., Asselineau, D., Dubetret, L., Coulomb, B., Lapiere, C., Nusgens, B., Neveux, K. The reconstitution of living skin. *Journal of Investigative Dermatology*, Vol. 81, no. 1, 1983.
85. Rao, K.P. Recent developments of collagen-based materials for medical applications and drug delivery systems. *Journal of Biomaterials Science*, Vol. 7, No.7, pp. 623–645, 1995.
86. Hokugo, A., Takamoto, T., Tabata, Y. Preparation of hybrid scaffold from fibrin and biodegradable polymer fiber. *Biomaterials*, Vol. 27, pp. 61–67, 2006.
87. Bensaid, W, Triffitt, J.T, Blanchat, C, Oudina, K, Sedel, L, Petite, H. A biodegradable fibrin scaffold for mesenchymal stem cell transplantation. *Biomaterials*, Vol. 24, pp. 2497–2502, 2003.
88. Jenks, M., Craig, J., Green, W., Hewitt, N., Arber, M., Sims, A. Tegaderm CHG IV securement dressing for central venous and arterial catheter insertion sites: A NICE medical technology guidance. *Applied Health Economics and Health Policy*, Vol. 14, No. 2, pp. 135–149, 2016.

89. Jabbari, E., Kim, D.H., Lee, L.P. *Handbook of Biomemetics and Bioinspiration, Bioinspired Materials*. World Scientific Publishing Co. Ptd. Ltd, 2014.
90. El Ghalbzouri, A., Lamme, E.N., Van Blitterswijk, C., Koopman, J., Ponec, M., *Biomaterials*, Vol. 25, No. 2987, 2004.
91. Aramwit, P. Introduction to biomaterials for wound healing. *Wound Healing Biomaterials* Vol. 2. 10.1016/B978-1-78242-456-7.00001-5.
92. Auger, F.A., Rouabhia, M., Goulet, F., Berthod, F., Moulin, V., Germain, L. Tissue-engineered human skin substitutes developed from collagen populated hydrated gels: clinical and fundamental applications. *Medical & Biological Engineering & Computing*, Vol. 36, pp. 801–812, 1998.
93. Wang, C.L., Miyata, T., Weksler, B., Rubin, A.L., Stenzel, K.H. Collagen-induced platelet aggregation and release. *Biochimica et Biophysica Acta*, Vol. 544, pp. 555–567, 1978.
94. Birla, R.K., Borschel, G.H., Dennis, R.G., Brown, D.L. Myocardial engineering in vivo: Formation and characterization of contractile, vascularized three-dimensional cardiac tissue. *Tissue Engineering*, Vol. 11, pp. 803–813, 2005.
95. Sahni, A., Francis, C.W. Vascular endothelial growth factor binds to fibrinogen and fibrin and stimulates endothelial cell proliferation, *Blood*, Vol. 96, pp. 3772–3778, 2000.
96. Rockwood, D.N., Akins, R.E., Parrag, I.C., Woodhouse, K.A., Rabolt, J.F., Culture on electrospun polyurethane scaffolds decreases atrial natriuretic peptide expression by cardiomyocytes in vitro, *Biomaterials*, Vol. 29, pp. 4783–479, 2008.
97. Lee, J.H., Kopecek, J., Andrade, J.D. Protein resistant surfaces prepared by PE containing block copolymer surfactants. *Journal of Biomedical Materials Research*, Vol. 23, pp. 351–368, 1989.
98. Hill-West, J.L., Chowdhury, S.M., Slepian, M.J., Hubbell, J.A. Inhibition of thrombosis and intimal thickening by in situ photopolymerization of thin hydrogel barriers. *Proceedings of the National Academy of Sciences of the USA*, Vol. 91, pp. 5967–5971, 1994.

3 Amorphous Semiconductors
Past, Present, and Future

Neeraj Mehta
Banaras Hindu University

CONTENTS

3.1 Background of Problem 38
    3.1.1 Distinction between Crystalline and Amorphous Semiconductors ...40
    3.1.2 Analogy between Amorphous and Crystalline Materials 42
    3.1.3 Techniques Used to Distinguish between Amorphous and
        Crystalline Materials 43
    3.1.4 Classifications of Amorphous Semiconductors 44
        3.1.4.1 Covalent Amorphous Semiconductors 45
        3.1.4.2 Ionic Amorphous Solids 46
        3.1.4.3 Metallic Amorphous Solids 46
3.2 Band Models for Amorphous Semiconductors 47
    3.2.1 Cohen-Fritzsche-Ovshinsky (CFO) Model 48
    3.2.2 Davis and Mott Model 48
    3.2.3 Mott, Davis and Street Model 48
3.3 Preparation of Amorphous Semiconductors 51
    3.3.1 Quenching Technique 51
    3.3.2 Thermal Evaporation Technique 51
    3.3.3 Flash Evaporation Technique 52
    3.3.4 Sputtering Technique 52
    3.3.5 Glow Discharge Decomposition Technique 53
    3.3.6 Chemical Vapor Deposition Technique 53
    3.3.7 Pulsed Laser Deposition 53
    3.3.8 Other Techniques 53
3.4 Experimental Techniques to Study Amorphous Materials 54
    3.4.1 Electrical Characterization 54
        3.4.1.1 DC Conductivity Measurements 54
        3.4.1.2 AC Conductivity Measurements 54
        3.4.1.3 Defect State Measurements 55
    3.4.2 Optical Characterization 55
3.5 Applications of Amorphous Semiconductors 59

| 38 | Functional and Smart Materials |

3.6 Present Status 60
    3.6.1 Our Understanding in This Area 60
    3.6.2 Problems for Further Research in This Area 61
Acknowledgments..62
References..62

3.1 BACKGROUND OF PROBLEM

One of the fundamental and the most interesting topics in engineering and science is the distinct states of the ordered and disordered materials. The absence of Long range order (LRO) is the essential aspect that differentiates an amorphous solid with its crystalline counterpart. There is no translational periodicity in amorphous solids. The unpredictability at large separations occurs in these disordered materials as signature of the absence of long-range ordering. About an atom under consideration, the atomic scale structure is extremely non-random for a small number of inter-atomic distances. In the structure of amorphous solids, the short-range order (SRO) is immensely treated as the signature of disordered state. A glass is one of the members of the family of non-crystalline solids that are synthesized by the traditional route of melt-quenching. The non-crystalline materials prepared by unusual synthesis techniques (such as sol-gel method, solid-state amorphization processes and thermal evaporation) are occasionally defined amorphous solids. Thus, it is very tough to extort any significant information about the structural units by performing the diffraction experiments on such disordered materials due to the presence of only SRO.

The scientists working in the field of the amorphous semiconductors have long been concerned by their structural, optical, and electrical properties. Even their extreme existence has led to elementary research problems, and their theoretical investigations have been delayed by the entire disintegration of straightforward techniques. On the other hand, the observations of amorphous solids indicate that the spectrum corresponding to their electrical and optical properties is every bit as diverse as that of crystalline materials, amorphous metals, semiconductors, and insulators are all set up in the natural world. In the last six decades, enormous development has been made in the theoretical analysis of structural properties of amorphous semiconductors. Such studies have revealed an immense conception too about the structure of crystalline units. At this juncture, it clearly seems that we possess a fundamental perceptive of the qualitative characteristics of the amorphous solids. Though, the scrupulous theory and the quantitative mechanism are still lacking for the preceding forecast.

When we are dedicated to sacrifice the inferences of the periodicity corresponding to the long-range, we can make the growth of a structure for the study of amorphous solids equivalent to that of their crystalline counterparts. In reality, the Hamiltonians for the many body problems of both amorphous and crystalline phases are indistinguishable. There are three necessary components that are involved in this parameter. The first two components are the kinetic energies of the ion cores and the outer electrons while the overall electrostatic interactions between the charged particles play the role of the third component. The basis that solid materials preserve their shape in free space necessitates an equilibrium configuration for the structure having no internal forces between the ion cores. This structure is always determined empirically, in general with the aid of diffraction or absorption experiments, and it is an essential

Amorphous Semiconductors

preliminary point for any additional analysis of the properties of the material. Like crystalline materials, the periodicity of the structure in amorphous materials does not exhibit the long-range order. Thus, the presence of SRO in the amorphous structure is the distinctive characteristic that generally differentiates the amorphous phase from the crystalline phase. However, the chemistry of the basic elements in both phases is reflected by their corresponding structures. Thus, there is similarity in the nearest-neighbor surroundings of any particular atom for the equivalent crystalline and amorphous solids in the majority of case studies. The appropriate point of view in our current outlook is that the local optimization of the chemical bonding in any alloy is achieved because of its solidification in such a way that it can provide the favorable situation. A local minimum in potential energy results as a general rule in the disordered structures having a periodic structure that can characterize the absolute minimum in energy for a given assortment of the constituent atoms and such a structure consists of a significant subclass of solid materials.

Extensively, it was supposed that the synthesis of the amorphous solids is only possible by employing a restricted number of materials. Consequently, some special amorphous solids such as organic polymers and glasses were frequently referred to as the "glass-forming solids." In fact, this conception was incorrect, and later it is found that the tendency of the glass formation is almost a common feature of the condensed matter. In 1955, B. T. Kolomiets and his associates in the USSR first tried other VI group elements (S, Se, and Te) in place of conventional element "O" which was used to prepare oxide glasses [1]. They found that glasses prepared using one and/or more chalcogen elements (i.e., sulfur, selenium, and tellurium of the VI column of the periodic table) showed the semiconducting properties. A special name "chalcogenide glasses" was set to these glasses to differentiate them from oxide glasses identified at that time. Chalcogenide glasses are defined as the molecular amorphous solids which have the chalcogen elements in the prominent quantity. An extensive research on amorphous semiconductors started when S. R. Ovshinsky reported the different kinds of switching phenomena that characterized a huge category of amorphous solids, and the subsequent promotion unfolding several potential applications for these phenomena [2–4]. The utilization of chalcogenide thin films for the memory of the computer was also demonstrated by him. Subsequently, a fabrication technique for n-type and p-type doping of amorphous silicon was developed by the Spear and LeComber. These two British researchers performed this work in the University of Dundee, U.K., and reported that the approach developed by them was nearly comparable to the existing technique for the crystalline complement. Since then, the amorphous semiconductors have drawn the attention of various research groups all over the world for their utilization in various solid-state devices.

As contrasting to a crystalline semiconductor, an amorphous semiconductor possesses a structural network having short-range periodicity. Prior to establishing the significance of amorphous materials, the basic concepts of the quantum picture of solids were already built up. This makes possible the simplification of complicated calculations of their physical properties by using the symmetric features related to the periodicity. Consequently, a fast perceptive of the common character and the properties of amorphous solids and primary grounds for such features could be established keeping in mind the existing knowledge of crystal engineering and physics. Even after the discovery and the characterization of the amorphous semiconductors, an incredibly huge attempt was exhausted into understanding the properties of those amorphous materials with straightforward crystalline counterparts. The purpose of present chapter is to highlight the overview of amorphous semiconductors, some of their physical properties and applications.

3.1.1 Distinction between Crystalline and Amorphous Semiconductors

Traditionally, the crystalline solids have overstated importance because of (i) the experimental determination of their structure is much easier; (ii) the quantitative analysis of their electronic properties are much more agreeable; and (iii) the decoupling of the specific parameters by employing the existence of crystalline restrictions in a controlled way like the careful chemical alloying. Thus, the doping of a semiconductor (e.g., Si) is possible by using the small amount of specific dopants (e.g., phosphorus or boron). The overall composition of a solid controls its optical properties, while small defect densities present in the solid play a crucial role in controlling its electrical properties. Thus, the doping preserves the huge optical energy gap of silicon while it causes a minute reduction in the electronic gap.

Since the capability to dope helps in the improvement of solid-state devices (e.g., transistors and lasers), doping has drawn the great attention of engineers, technologists and physicists to improve the properties of the crystalline semiconductors.

However, a great deal of the current insights into amorphous semiconductors has made possible that (i) the amorphous solids show a numerous wider range of behavior due to the deficiency of periodic constraints; (ii) the additional alteration in the structure can be done by using the approach of non-equilibrium synthesis; and (iii) if the suitable deposition circumstances are conquered, the conversion of roughly every alloy in the shape of thin film is achievable (irrespective of its solubility).

Unlike, crystalline semiconductors, the amorphous semiconductors have a random arrangement of atoms. So they possess SRO of few atomic distances but have complete absence of LRO periodicity of atoms [5]. This fundamental difference is shown in Figure 3.1. Crystalline semiconductors have chemical bonds of definite lengths and angles. However, in amorphous semiconductors, chemical bonds and angles may vary. All chemical bonds in crystalline semiconductors are chemically satisfied, whereas in amorphous semiconductors, all chemical bonds are not chemically satisfied and known as "dangling bonds." Crystalline semiconductors are anisotropic in

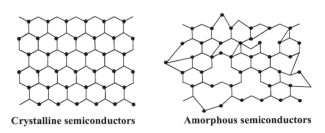

FIGURE 3.1 Structure of crystalline and amorphous semiconductors.

Amorphous Semiconductors

41

nature, i.e., the properties in different directions differ considerably, whereas amorphous semiconductors are isotropic in nature, i.e., the properties do not depend upon the direction of measurements.

There is no way to confirm the crystalline or non-crystalline nature of the material by its external appearances alone. With the discovery of X-ray diffraction only, it could be ascertained whether a particular substance is crystalline or non-crystalline. Since the theoretical analysis of semiconductors was also established on the concept of the periodicity of atoms in solids, it was prolonged idea that non-crystalline solids could not be semiconductors. No experimental result was also available which showed the semi-conducting behavior of glassy substances. The list of some exceptional characteristics of amorphous semiconductors that distinguish them with their crystalline counterparts are as follows:

- In amorphous semiconductors, thermodynamic equilibrium is not a prime requirement. In other words, such equilibrium does not exist in amorphous semiconductors. These semiconductors are stable due to the unavailability of sufficient thermal energy at room temperature for overcoming the potential barrier, and consequently, the structural transformation of such semiconductors from amorphous to crystalline state is usually not possible at lower temperatures. However, when we heat an amorphous semiconductor at higher temperatures, it transforms from amorphous to crystalline state through a thermally induced exothermic phase transition (i.e., we observed a release of considerable energy). As already mentioned that thermodynamic equilibrium is not necessary, the amorphous semiconductors may survive in numerous diverse structural states. Consequently, alloying can be done by using two or more elements of such semiconductors, i.e., they can be made in various compositions. The composition of amorphous semiconductors, therefore, becomes a convenient way to vary their properties by varying the composition. Hence, the amorphous semiconductors are also known as tailor-made materials. This type of variation is not possible in crystalline semiconductors due to the requirement of thermodynamic equilibrium to maintain the minimum energy state.
- The SRO of amorphous semiconductors can be changed by external means. Thus, the properties of these semiconductors can be changed drastically through local structural transformation by exterior perturbations (e.g., light exposure, high electric field, etc.). The structure becomes a new variable that can be used for making solid-state devices using these semiconductors.
- A crystalline semiconductor, when irradiated with high energy particles, tends towards non-crystalline state. The properties change drastically. On the other hand, radiation does not affect amorphous semiconductors much as they are already in non-crystalline state. This is probably the main reason why amorphous semiconductors are explored with great enthusiasm so that they could be used in space applications where crystalline devices produce noise due to radiation.
- The technology of preparing amorphous semiconductors is much easier and therefore less costly than growing single crystals for crystalline devices.

Crystalline materials are generally used for devices in single crystal form, which requires quite sophisticated technology as compared to amorphous semiconductors. In other words, the conversion of the amorphous semiconductors in the shape of thin films can easily be achieved by sputtering technique or vacuum evaporation. Due to their conversion in thin-film form, integrated circuits can easily be fabricated. All these simplifications have made amorphous semiconductors most suitable for cheaper semiconducting technology.

3.1.2 Analogy between Amorphous and Crystalline Materials

Our information about the structural properties of the amorphous solids appears partial when we compare them to the structural properties of crystalline solids. Specifically, there are still lacking the lots of concepts about the idea of a perfect amorphous solid and the corresponding theory. According to recent concepts of modern physics, we can illustrate an amorphous solid as a disordered crystalline solid having some degree of order midway a perfect crystalline solid and a perfect amorphous solid. This idea is to stand on the basis that glasses are independent of any such restriction that guides the organization of the atomic clusters in the crystalline solids. Consequently, there is an extent of vagueness in the approach that is employed for the positioning and orientation of adjacent clusters.

A promising inference developing from this outlook is that amorphous phases (i.e., disordered states) are derived from the consequent crystalline phases (i.e., ordered states). In the field of geometrical crystal physics, we imagine the existence of defects for defining a disordered crystalline structural state with respect to the absolutely ordered structural state. Thus, we can put forward the idea about the disordered materials as the crystalline materials that can be rebuilt their absolutely ordered structural state by the setback of defects. However, this idea of geometrical crystal physics cannot be applied directly for the amorphous semiconductors. In other words, we can promote this concept by keeping in mind their haphazard atomic orientations. This can be understood by using the illustration shown in Figure 3.2.

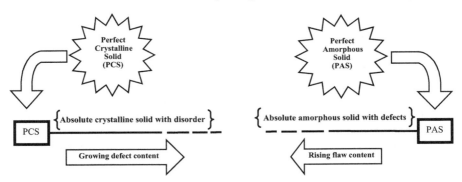

FIGURE 3.2 Overview of the structure of solids along an indeterminate and a random variable to some extent. The squares designate the locations of the perfect or ideal structures, while the lines point out the expansion of the structures in the absolute solids.

Amorphous Semiconductors   **43**

This figure shows the template of comparative sites of the two kinds of solids and the prediction about the discontinuation between the corresponding two kinds of atomic orientations (i.e., disordered or orders states). The extreme minute gap between the square and the line on the left-hand side (i.e., the crystalline side) depicts the possibility of grown of the nearly ideal single crystal. The larger gap on the right-hand side (i.e., the amorphous side) shows that the structural properties of glasses may not be as near to that of the perfect amorphous solid. A discontinuation in the line close to the center is intended to designate that even extremely disordered crystalline solids are not identical as extremely flawed amorphous glasses, and vice versa. Similar concept was proposed by Tanaka et al. [6].

3.1.3 TECHNIQUES USED TO DISTINGUISH BETWEEN AMORPHOUS AND CRYSTALLINE MATERIALS

Discrimination between crystalline and non-crystalline materials is necessary to find out whether a given specimen is amorphous beyond any doubt, i.e., absolutely destitution of the crystallinity at micro-scale or nano-scale level. Perceptibly, the primary preference in this regard is the diffraction phenomenon in which the bulk sample plays the role of diffraction element to diffract the X-rays or neutrons. If the shape of the diffraction element (i.e., sample) is a thin film, then we chose the electrons for the diffraction experiment. However, the existence of sharp Bragg-like spots placed over the diffuse haloes can be noticed occasionally as the signature of insignificant crystallinity in the transmission-electron micrographs of some specific amorphous materials [5].

Consistently, the dependence of the scattering intensity I on the scattering vector $\vec{K}$ of the scattered radiation can be used to mark the crystallinity, where the magnitude of the scattering vector $\vec{K}$ can be expressed in terms of scattering angle θ and the wavelength λ as:

$$\left|\vec{K}\right| = \frac{4\pi\sin\theta}{\lambda} \tag{3.1}$$

The radial distribution function (RDF) is another parameter to mark the order of crystallinity. RDF is generally defined as the possibility of finding an atom in the range between the distances r and $(r+dr)$ from an atom that is treated as the origin. The peaks in the RDF linked with the crystallites, at the distances away from that equivalent to the first peak, are considerably lesser than those features of the amorphous solids. The effects due to thermal and instrumental broadening play the crucial role in the determination of each peak width in the RDF of the crystalline structures. Conversely, for amorphous materials, the topological disorder leads to an increasingly bigger expansion of the peaks with raising the distance r. Such peaks amalgamate into the unremarkable (parabolic) curve, which is distinctive of the RDF of the entire structural disordered state of matter. In specific cases of the iso-structural materials (such as Si and Ge), the occurrence of the peaks in the crystalline RDF is observed at such distances where the peaks are absent for their amorphous counterparts. This point can be understood by an example of crystalline Germanium (c-Ge). The c-Ge has the peak at position 4.7 Å, which arises from cross-ring relationships for the rings having six members of a particular conformation initiated in the diamond-cubic structure of c-Si or Ge, and very few such rings are found in the equivalent amorphous structures.

Let us consider the Debye–Scherrer relation for the determination of the crystallite size D:

$$D = \frac{\lambda}{\Delta Q . \cos \theta} \tag{3.2}$$

Here ΔQ is the width of the peaks in the scattering pattern. From this relation, it is clear that the width ΔQ is decreased with the increase in the crystallite size D.

The widths of the peaks can be equivalent to those representatives of the amorphous materials for the crystallite sizes of a few angstroms. Thus, the diffraction is no longer able to discriminate between amorphous and (nano) crystalline materials in this frontier. When diffraction is no longer helpful, the calorimetric experiments may be able to be used to differentiate between amorphous and nano-crystalline or micro-crystalline materials. Here, the critical tip is that the material changes by a process of growth in grain size with heating if the material is already in the crystalline phase. However, if the sample is an amorphous solid in the beginning, the material transforms by a method of crystallite nucleation and subsequent grain growth. These two methods are distinct from each other in terms of the isothermal rate of change of enthalpy release, which is analyzed as a function of annealing time [7].

3.1.4 Classifications of Amorphous Semiconductors

Figure 3.3a illustrates a landscape of the different types of amorphous materials as a plot where chemical bonding nature is taken on x-axis and the band gap is taken on y-axis axis. We can effortlessly perceive from the figure that the chemical bonds in the usual amorphous materials show the high degree of covalent character while the wide-gap materials (i.e., transparent materials) have insulating behavior from the electrical point of view.

The electrical characterization of the amorphous materials reveals that they fall in the categories of both insulators and semiconductors as well as in the category of the conductors in some specific situations. Some amorphous materials demonstrate superconductive behavior too at exceptionally low down temperatures. Semiconducting properties of the amorphous semiconductors have been observed in the single elements, compounds as well as in multi-component alloys. Generally, almost all kinds of bonding (i.e., ionic, covalent, metallic, Vander-Waals or hydrogen bonding) may be observed in the different types of amorphous semiconductors [8]. Further, the subdivision of such disordered materials into diverse categories depends on their physicochemical properties and the approach employed for their synthesis. Broadly, covalent amorphous semiconductors, ionic solids (oxide semiconductors) and metallic amorphous semiconductors are the three main groups that are used for their classification. The details of this classification are described in the subsections.

Amorphous Semiconductors

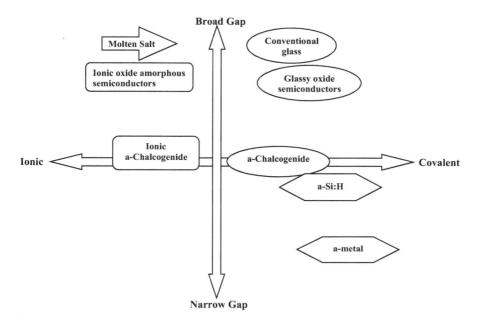

FIGURE 3.3A Site-map of amorphous materials (here "a-" stands for amorphous).

3.1.4.1 Covalent Amorphous Semiconductors

These types of amorphous semiconductors can be further separated into the following three subcategories:

3.1.4.1.1 Tetrahedral-Bonded Amorphous Semiconductors (TAS) Films

This subcategory consists of elemental materials such as Si, Ge (Group IV elements) are III–V semiconductors. Various deposition techniques are used for their synthesis in the form of thin films. It is not possible to prepare this class of materials in the amorphous form by employing the melt-quench technique due to the highly coordinated structural characteristics of the melt.

3.1.4.1.2 Chalcogenide Glassy Semiconductors (CGS)

Chalcogenide glassy semiconductors (CGS) are also known as two-fold coordinated amorphous semiconductors. The chalcogen elements (S, Se, Te) are the main components of the alloys lying in this subcategory, and their composition is high in the alloys as compared to other constituent elements. The presence of two non-bonding p-orbitals in two-fold coordination is responsible for their lot of distinctive properties, and so this is the unique characteristic of such glassy semiconductors. Due to this feature, they are widely known as "lone-pair semiconductors." The value of electrical conductivity in CGS lies in the range 10^{-13} to 10^{-3} (Ω cm)$^{-1}$. The initial CGSs are identified as elemental semiconductors such as S, Se while the binary alloys or compounds (e.g., As_2S_3, As_2Se_3, etc.) belong to the first generation of CGS.

There is significant ordering in terms of chain or layered structure in the elemental CGS and their first generation that widens locally in 1D or 2D networking. The 3D networking starts in further generations like ternary and quaternary alloys. The understanding of electronic states in chalcogenide glasses makes them suitable for several applications.

The intrinsic semiconducting character of CGS was observed by Kolomiets [9] when he found that there is no significant variation in the electrical conductivity after the incorporation of certain dopants. The research group of Mott [10,11] reported that it is not only the disorderness in these materials that is responsible for the localized states in CGC but also the presence of distinctive defects states that play an effective role in this class of materials. By proposing the conception of dangling bonds, they successfully and satisfactorily elucidated the exceptional properties of CGS.

3.1.4.1.3 Tetrahedral Glasses

The general formula to express this class of materials can be expressed as $A^{II}B^{IV}X_2^{V}$. Here the upper suffix identifies the corresponding group of the periodic table. The location of the Fermi energy level close to the center of the band gap is noted as remarkable characteristic of the tetrahedral glasses. But still, the glasses of this class are either n-type semiconductors or p-type semiconductors.

3.1.4.2 Ionic Amorphous Solids

The materials of this class are defined as the inorganic glasses that consist of the combination of the silicates having the well-built ionic bonds. It is interesting to note that they show the character resembling with that semiconducting oxide glasses. The window glasses, glass tubing and optical instrumentation are some applications of these glasses that are extensively used in daily life. The vanadium phosphates, halide/oxide glasses, and iron phosphate glasses are some examples of this class. There is tight bonding of the electrons with their corresponding ions. Consequently, the electrons do not play an effective role in the conduction mechanism of electrical conduction. In other words, these materials behave like good quality insulators. The members of this class possess some semiconducting properties too due to the presence of the transition metal ions in valence band. The transfer of the charge between the transition metal ions or hopping of the charge carriers from one localized state to the other are the two possible conduction mechanisms.

3.1.4.3 Metallic Amorphous Solids

Metallic glassy alloys are the well-known members of this class of materials that consist of some base metals (e.g., Fe, Ni, Al, Co, Mn, Cr, and Cu) collectively with economical metalloid (e.g., B, C, Si, P, N, Ge and As). Metallic glasses that are the alloys of Fe, Ni and Co behave like the amorphous solids. Additionally, GdCo, GdFe, and GdTbFe are also the members of this class having rare-earth transition elements, and they demonstrate the high-quality magnetic properties that better than their crystalline complements [12].

Figure 3.3b demonstrates the flow chart of the aforesaid classification of amorphous materials.

Amorphous Semiconductors 47

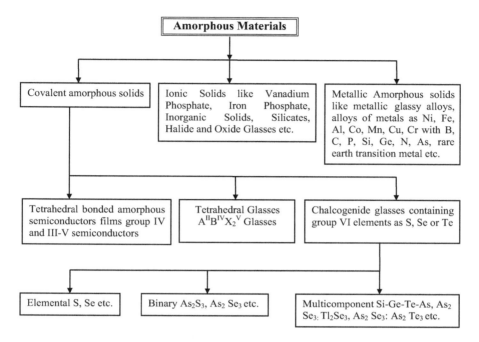

FIGURE 3.3B Classification of amorphous materials.

3.2 BAND MODELS FOR AMORPHOUS SEMICONDUCTORS

The knowledge of energy band structure is needed to understand the electrical behavior of amorphous semiconductors. In crystalline semiconductors, the density of states (DOS) possesses van Hove singularities and unexpected terminations at the edge of the valence and the conduction bands. Thus, the presence of sharp edges in the DOS provides a basis for a distinctive forbidden energy gap. The extension of such states takes place within the range of the band as a signature of occupancy of the wave functions over the whole volume. The local structural environments in crystalline and amorphous semiconductors are found to be nearly the same. Hence, the distribution of the density of energy states is similar in crystalline and amorphous semiconductors. The effect of disorder is to introduce band tailing by removing van Hove singularities [13,14]. The elimination of van Hove singularities is one of the consequences of the disorderness on the DOS that play a significant role in the smoothening of the structural sharpness in the distribution of DOS. This is responsible for the tailing of states into the gap. An even more important consequence of disorder is that the electron states become "localized" which means that the electron present in such state would be spatially confined to the vicinity of predominately a single atomic site. Further, the presence of such localized states corresponding to tail states is a consequence of the disorder.

Because of the great variation in character, properties and constituents of amorphous semiconductors, various models and their modifications have been proposed to provide a satisfactory explanation for the perceptive of the essential characteristics of these semiconductors. The arbitrariness behind the arrangements of atoms in amorphous systems influences sturdily the mobility of the carriers. According to Mott [15], the localized states in the amorphous materials are probably created when the spatial fluctuations occurs in the potential due to the configurationally disordered state. A band gap may, however, exist in amorphous semiconductors due to SRO. The absence of sharp cut off in the valence and conduction bands is due to localized states, and therefore, they have only a tail beyond and underneath the normal band. The enhancement in the degree of such tailing is anticipated in the alloys having both the compositional and the positional disordered networks. To describe the characteristics of the amorphous materials, three basic models have been proposed by various workers, which are discussed below.

3.2.1 Cohen-Fritzsche-Ovshinsky (CFO) Model

The above model has been proposed by Cohen, Fritzsche, and Ovshinsky [16]. They suggested that in case of disordered materials such as chalcogenide glasses, the mutual overlapping of the valence and conduction bands tail takes place near the center of the forbidden gap. The mobility of carriers has finite values in high-density states but it reduces unexpectedly in the tail states. The term "mobility edges" is frequently used to express such boundaries. The critical energies at mobility edges define mobility gap. The features of CFO model discussed above are shown in Figure 3.4 a and b.

3.2.2 Davis and Mott Model

The modeling of Davis–Mott [17] states that when we consider the saturation of all bonds inside a perfect amorphous semiconductor and the complete nonexistence of the long-range fluctuations, then we can expect a real band gap having a DOS as depicted in Figure 3.4. Further, this model [17] suggests the pinning of the Fermi level between the two tails as a characteristic of a fine band of compensated localized states as shown in Figure 3.5. In this figure, E_C and E_V show the position of the mobility edges for electrons and holes corresponding to conduction and valence bands. Thus, the probability of the hopping conduction in the localized states is another important point of this model.

The existence of the deep tails is found only with the fluctuations in the density or bond angle. However, Cohen et al. [16] proposed the probability of the strong overlapping of the tails in some amorphous liquids. The expended fluid of mercury is an example of such amorphous liquids. When there is a significant variation in the volume and temperature of the expended fluid of mercury, then it experiences a metal to insulator transition.

3.2.3 Mott, Davis and Street Model

The famous model of Mott, Davis, and Street, i.e., MDS model is frequently used for the explanation of the general features of amorphous semiconductors [10,11,17]. The position of the mobility edges is same in this model like in the previous model. The nonexistence of long-range order is taken as a basis to explain the dissimilarities

Amorphous Semiconductors

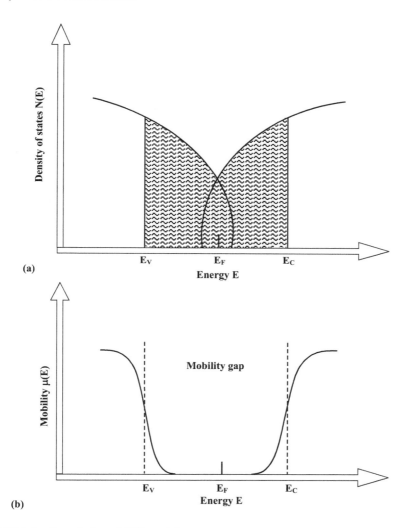

FIGURE 3.4 Description of CFO model.

between the localized states and the defects in structure. According to MDS model, the defect states built the longer tails having inadequate density, and such tails are responsible for the pinning of the Fermi level in the vicinity of the center of the gap. According to Mott [18,19], there is strong possibility of the splitting of this center band into two levels, which are known as the donor and the acceptor band as shown in Figure 3.6. The states close to the Fermi level are, then, considered because of the presence of defects (especially the dangling bonds are taken into account). The MDS model was inspired by the Anderson's idea [20] of the pairing of the electronic states, which was proposed by Anderson preferably for the CGS. The breaking of the bonds occurs in some structural configurations because of the incapability of sharing the electrons between the atoms. Hence, a dangling bond emerges out of this

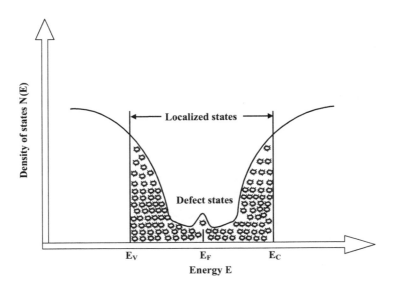

FIGURE 3.5 Description of Davis and Mott model.

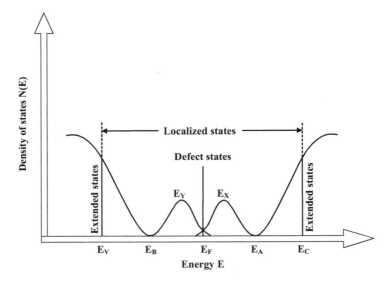

FIGURE 3.6 Description of Mott, Davis, and Street model.

unstable situation. A singly occupied dangling bond is described by the neutral state (i.e, D^0), thereby indicating the prediction of a signal due to the electron spin resonance (ESR). In some cases, a dangling bond can attract an additional electron and, thus providing a lone pair. Consequently, we do not observe any ESR signal and then we designate such dangling bond by a negative state (i.e., D^-). The corresponding positive state (i.e, D^+) represents the dangling bond when the bond loses even when it has a single electron. In such situation, a bond without any spin is appeared as a

Amorphous Semiconductors 51

signature of a hole. The MDS model envisages the nonappearance of ESR signal and the para-magnetism, and it is capable of explaining a lot of other properties of amorphous semiconductors.

3.3 PREPARATION OF AMORPHOUS SEMICONDUCTORS

Amorphous semiconductors are synthesized by means of non-equilibrium routes. Several synthesis techniques are feasible, depending on what kinds of materials are required for the research and/or application. For the synthesis of so-called glasses and glass fibers, the quenching from the liquid state is a widely accepted way (also known as melt-quenching). When there is a need for the formation of the samples in thin-film geometry, then we employ the techniques such as the thermal evaporation, sputtering, and chemical vapor deposition (CVD). The production of the amorphous materials is also possible by using crystalline solids as targets and targeting them by ion bombardment and highly intense electromagnetic irradiation.

The details of the aforesaid preparation techniques of amorphous semiconductors are as follows [21]:

3.3.1 Quenching Technique

In this technique, materials are prepared in amorphous form by rapid cooling from the melt. Such materials are called glassy (a name given to those solids, which are made non-crystalline by swift cooling from the melt). For the synthesis of glasses, weighing the constituent elements in appropriate atomic weight percentages is an essential and initial step to achieve the desired composition before sealing the sample inside a quartz tube (i.e., ampoules). This ampoule is reserved in a furnace at an appropriate high temperature so that all the components get melted. The sealed ampoules are frequently rocked for about 20 h to ensure the homogeneity of the melt [22].

The quenching is the final step that is completed by dropping the quartz ampoules abruptly in a tub that is filled either with ice-cooled water or liquid nitrogen. The choice of chilled water or liquid nitrogen depends upon the requirement. In some cases, air quenching is adequate where air is blown on the heated quartz tube by an air blower. It is important to note that melt-quench is not universal approach for the glassification of all kinds of materials (e.g., silicon and germanium), no matter how much fast the rate of cooling. When we want to convert an alloy in glassy form, then it is possible only for some range of the definite amounts of constituent elements. An organized analysis is necessary to discover an appropriate phase diagram for the glassification of a certain series of alloys. This is done by the preparation of each sample in a small amount and then the glassy character is ascertained by X-ray diffraction technique.

3.3.2 Thermal Evaporation Technique

To synthesize the materials in the amorphous phase, this approach is one of the most straightforward methods. In this technique, firstly we evacuate the given sample by creating a high vacuum inside a bell jar where the sample is enclosed on a filament for the transfer in the amorphous phase. Next, we heat this filament by using the concept of Joule-heating so that it can be converted in the form of thin films [23]. For the proper thermal evaporation, the preliminary materials are converted in the form of ingots or powders before mounting them in a specially designed sample boat. The Joule-heating causes the melting followed by thermal evaporation of the sample in the chamber having vacuum at around 10^{-6} Torr (~ 10^{-4} Pa). To attain such a high vacuum, a combination of a rotatory and an oil-diffusion pump is employed along with a liquid-nitrogen-cooled trap. In most of the coating units, the electrical heating (i.e., Joule-heating) is performed by the boat itself. Thus, we choose the elements having extremely high melting points, such as molybdenum (Mo) or tungsten (W), to make the boats.

The underlying substance, which is used for deposition, is usually made of glass or any other appropriate material. It is widely known as substrate, and a heater is escalated inside the bell jar to control its temperature. There are different kinds of typical coating units that are accessible in the industries for manufacturing of thin films by employing thermal evaporation. Generally, this method is appropriate for the deposition of the materials having low melting point by the condensation of their corresponding vapors on the preferred substrates. We can control the deposition rate and the thickness of thin films by varying the temperature of the boat and the deposition time, respectively.

3.3.3 FLASH EVAPORATION TECHNIQUE

The approach used in this technique is almost identical to the previous technique excluding that the sample is sprinkled from a magnetic strip on the boat which is already heated [21]. The fractionation effect is reduced in this technique. In addition to the coating unit used above, an AC magnetic field is set up to create vibration in the magnetic strip.

3.3.4 SPUTTERING TECHNIQUE

In sputtering, the erosion of the sample is achieved either atom by atom or by clusters of atoms due to the bombardment of the energetic ions on the target. This ionic irradiation is produced by using low-pressure plasma. Consequently, deposition of thin film takes place on the substrates. The most convenient manner to stimulate sputtering is to apply a high negative voltage across the surface of the target so that the positive ions can be attracted from the plasma. However, the method of DC sputtering is realistic merely when the target is adequately conducting so that target can play the role of the electrode.

The mechanism of the sputtering is more complex than that of thermal evaporation. Thus, we must be aware about various parameters that control the sputtering. The radio-frequency (RF) power applied to the target and the bias voltage of the substrate are two examples of such parameter. It is interesting to note that when we want to prepare thin films of a multi-component system, then sputtering is undoubtedly better and beneficial as compared to evaporation because the composition of the target sample is approximately unchanged after their conversion in the resulting films. This is possible because the sputtering rate is independent of the constituent elements.

Amorphous Semiconductors

In case of insulating materials, a common approach is to apply RF field to the target. To accomplish this purpose, the RF voltage is capacitively coupled to the target surface. For metallic targets, this is achieved by inserting a capacitor in series with the target. A metal backing electrode plays the role of the capacitive component, and it is normally bonded to insulating targets. The material deposited on the substrate can form amorphous film for the same reason as in evaporation.

3.3.5 Glow Discharge Decomposition Technique

Like the previous method, the working principle of this method also depends on the creation of plasma in a low-pressure gas. But a self-chemical decomposition of the gas occurs rather than ions from the plasma ejecting material. This causes the deposition of a solid film on a substrate which is reserved in the plasma. An RF field is employed for the creation of the plasma either by using the inductive or capacitive coupling. The shape and size of the resulting films depend upon the rate of flow, gas pressure, temperature of the substrate, and chamber geometry. This technique is widely accepted all over the world to synthesize the hydrogenated amorphous silicon (a-Si:H), where substrate temperature is kept high to get good quality films.

3.3.6 Chemical Vapor Deposition Technique

The working principle of CVD is similar to the previous method due to dependence of both methods on the disintegration of the vaporized samples. The distinct feature of this approach is that it depends efficiently on the thermal energy for the disintegration, and the substrate is heated by means of the applied RF field so that the vapor can decompose on it.

3.3.7 Pulsed Laser Deposition

Pulsed laser deposition (PLD) is very simple when compared with sputtering and the various CVD techniques described so far. The sample (i.e., target material) is enclosed in a vacuum chamber and its evaporation is achieved by focusing a pulsed laser beam of high power. This resembles electron-beam evaporation, but a directional plasma cloud is produced in PLD due to the absorption of photons. Thus, its working principle is significantly different from that of the straightforward thermal evaporation [24].

3.3.8 Other Techniques

Apart from the above mentioned techniques, some other techniques such as "gel desiccation," "electrolytic deposition," "chemical reaction," and "irradiation" have also been used to prepare amorphous semiconductors in some special cases.

Mechanical milling (MM) [25,26] in solid state is another special preparation method that is not popular in this field. The principle behind the swift quenching method is the fast elimination of kinetic energy in an energized state. MM comprises a chemical reaction in a system with a negative heat of assimilation followed by

54   Functional and Smart Materials

homogenization procedure up to the atomic level. It proceeds through the annihilation of the crystal structure of an intermetallic compound or pure element because of the gathering of strains and defects. Before the beginning of MM, the starting material is vigorously in the thermodynamic equilibrium (i.e., crystalline) state. Therefore, the "uphill" process should be involved in the initial stage, during which energy is given mechanically to the material system. Firstly, the strain destroys the chain molecules and then the stimulation of the defects takes place.

3.4 EXPERIMENTAL TECHNIQUES TO STUDY AMORPHOUS MATERIALS

All the techniques used for the study of crystalline semiconductors are also being used to study amorphous semiconductors. Some important experimental techniques are given below.

3.4.1 ELECTRICAL CHARACTERIZATION

3.4.1.1 DC Conductivity Measurements

To understand the electrical transport, experiments on DC conductivity with temperature, Hall effect experiment, thermoelectric power with temperature have been used [27–29]. The Hall coefficient possesses a sign which is found to be opposite to what one expects from the concepts of crystalline semiconductors. It means that the n-type materials have positively signed Hall coefficient while p-type materials have negative signed Hall coefficient in amorphous semiconductors. This is explained in terms of conduction in localized states. However, the conductivity data in the same temperature range is explained in terms of the conduction mechanism applicable to the extended states. However, the correct sign of charge carriers in these materials is achieved in the thermoelectric power experiment.

3.4.1.2 AC Conductivity Measurements

AC conductivity measurements show that AC conduction occurs at audio frequencies start from low values (generally 100 Hz) in amorphous semiconductors, which is not observed in crystalline semiconductors in intrinsic state. Dielectric dispersion and dielectric loss is also observed in audio frequency range, which is also not observed in crystalline semiconductors at these frequencies. AC conductivity and dielectric data are explained in terms of defect states in the energy gap where hopping process is possible in the localized states [5].

Numerous amorphous semiconductors (e.g., chalcogenides and a-Si:H films) follow the following power-law dependence law [15]:

$$\sigma(\omega) = A\omega^s \tag{3.3}$$

This dependence of σ on ω is frequently known as "dispersive AC loss." This relation is generally valid for the angular frequency range 10^2–10^{10} Hz [15]. In Equation 3.3, the parameters A and s (<1.0) show the noticeable dependence on the temperature [30].

Amorphous Semiconductors

3.4.1.3 Defect State Measurements

The experiments for the determination of defect states include "field effect experiment," "space charge limited conduction measurements," "capacitance–voltage measurements," "photoconductivity measurements," "thermally stimulated current measurements," etc. These methods are same as applied to crystalline semiconductors [31–34]. The application of these methods to amorphous semiconductors requires some approximations, which must be taken very carefully. Results calculated from different experiments match reasonably well within experimental errors.

3.4.2 OPTICAL CHARACTERIZATION

For the optical characterization, several techniques are available in which the intensity of the light is measured after the reflection or transmission from the sample. In most of the cases, the samples are taken in thin-film form and the intensities are generally measured by varying the wavelength. Generally, the reflectance R varies with the variation in the path-difference Δ between the waves that are reflected from the air–film boundary. Similarly, the transmittance T shows the dependence on the path-difference between the waves that are reflected from the boundary of film-substrate. The value of Δ can be determined by using its relation with the incidence angle i, light wavelength λ and the thickness t of the film. Thus, the effect of change in the value of "t" is also an important study in addition to the variation in the value of λ [35]. The fast screening of the data is not achievable by such means in some cases when there is fast variation in the optical parameters but the thickness t varies gradually. To determine the optical parameters of the sample in thin-film form, various approaches have been designed. The general working principle of all such methods is to analyze the data of $R(\lambda)$ and $T(\lambda)$ [36,37].

The thin-film samples having low absorbance, slightly different approaches are used i.e., the determination of the optical parameters in such materials can be done by considering the envelopes of R versus λ plot and T versus λ plot. However, it is more suitable and straightforward to determine the plot of either R or T against λ. For this kind of samples, the techniques have been employed for the entire analysis of various optical parameters using reflectance measurements only. Unambiguously, the validity of the formulation for extrema in the values of R and T makes possible the determination of the optical constants by investigating their corresponding spectral dependence [38,39]. In the amorphous semiconductors, the short-range ordering and the defects linked with the amorphous network play a dominant effect on the optical absorption α and optical band-gap E_g. The optical characterization of the amorphous thin films of CGS can be done by the complete examination of various optical parameters.

The disorderness and the corresponding mechanism of defects interaction in amorphous thin films of CGS can be revealed by analyzing the variation of E_g and the short-range ordering as a function of thin-film thickness of a certain composition or the function of different compositions of same thin-film thickness [40]. The value of E_g is generally increased with the reduction in the disorderness and density of the defect states. A broad range of phenomena in amorphous semiconductors is covered by their optical characterization connected with the absorption phenomenon.

This assists us significantly in understanding the fundamental optical properties of amorphous semiconductors. Such optical characterization of the amorphous semiconducting thin films is also helpful in the advancement of optical instrumentation and device technology of the industrial research directly related to optics and optoelectronics. The experimental analysis of absorption spectra is the straight approach for inquiring the optical band gap of amorphous semiconductors and the corresponding structure. When the excitation of a photon of known energy (say $h\nu$) takes place between two energy levels, it provides a basis to study the corresponding absorption process. In a different approach, we can state that the absorption is also a consequence of the interaction of electromagnetic radiation with the atoms.

In semiconductors, a number of distinct optical absorption processes take place independently. The transition of electrons from the valence to the conduction band is one of such most significant processes (see Figure 3.7). This phenomenon is recognized as the basic absorption due to its significance. An electronic jump is observed as a signature of such a basic absorption, i.e., an upward transition (valence band → conduction band) of an electron by absorbing a light photon of the incident electromagnetic radiation. In such case, the minimum energy of photon, i.e., $h\nu$ is necessarily equal to E_g or larger than E_g. In other words, the following condition should be satisfied:

$$\nu \geq \frac{E_g}{h} \tag{3.4}$$

The lowest permissible value (i.e., E_g/h) of the photon frequency is denoted by ν_0, and it is treated as the measure of the absorption edge. This upward transition of electron by the absorption of a photon, the conservation laws of both momentum and energy are applicable. Mathematically, these laws can be expressed as:

$$E_{\text{final}} = E_{\text{intial}} + h\nu \tag{3.5a}$$

$$p_{\text{final}} = p_{\text{initial}} + \hbar k \tag{3.5b}$$

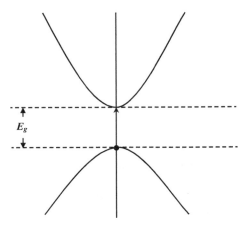

FIGURE 3.7 Fundamental absorption process in semiconductors.

Amorphous Semiconductors

The relations (3.5a) and (3.5b) clearly show that the electron–photon system obeys the conservation laws of both momentum and energy. In Equation 3.5a, E_{initial} and E_{final} are the electron energies before and after the electronic jump. In Equation 3.5b p_{initial} and p_{final} are the corresponding momenta of electron. The symbol "$\hbar$" is $(1/2\pi)$ times the Planck's constant while the symbol "k" represents the propagation constant of the photon absorbed in the process. Physically, it is the magnitude of wave vector $\vec{k}$ of the photon. In general, the value of k of a photon is extremely small in the visible region and so the Equation 3.5b reduces to $p_{\text{initial}} = p_{\text{final}}$. According to the selection rule, the transitions between the valence and conduction band in momentum-space are permitted merely in vertical directions (see Figure 3.7). Further, the quantum manipulations are necessary for determination of the absorption coefficient. Fundamentally, the incident radiation in the quantum approximations is considered as a perturbation. According to the *perturbation theory of the quantum* mechanics, this perturbation is responsible for the coupling between the electronic states of the valence and conduction bands.

At the energy gap, the absorption edge of crystal semiconductor ceases rapidly. On the other hand, the absorption of amorphous semiconductor does not show such termination of edge, and it expands into the energy gap region. The three researchers O' Leary, Johnson, and Lim proposed a model [41], which is known as OJL model. The OJL model correlates the optical absorption spectrum and the DOS to describe the distribution of electronic states for the amorphous semiconductors. According to OJL model, the distribution of DOS is governed by an exponential and a square-root functional dependence in the tail and the band regions, respectively, as follows:

$$N_c(E) = \frac{\sqrt{2}m_c^{*3/2}}{\pi^2\hbar^3}.f(E); \text{where } f(E) = \left\{ \begin{array}{l} \sqrt{E - E_c}, \, E \geq E_c + \dfrac{\gamma_c}{2} \\[2mm] e^{-1/2}\sqrt{\dfrac{\gamma_c}{2}}\exp\left(\dfrac{E - E_c}{\gamma_c}\right), \, E < E_c + \dfrac{\gamma_c}{2} \end{array} \right\}$$

(3.6)

$$N_v(E) = \frac{\sqrt{2}m_v^{*3/2}}{\pi^2\hbar^3}.f(E); \text{where } f(E) = \left\{ \begin{array}{l} \sqrt{E - E_v}, \, E < E_v - \dfrac{\gamma_v}{2} \\[2mm] e^{-1/2}\sqrt{\dfrac{\gamma_v}{2}}\exp\left(\dfrac{E - E_v}{\gamma_v}\right), \, E \geq E_v - \dfrac{\gamma_v}{2} \end{array} \right\}$$

(3.7)

Here E_c and E_v are the respective ground-level energies for the conduction and valence bands. In case of the amorphous semiconductors, the letters "v" and "c" denote the corresponding tail states. The parameters γ_c and γ_v signify the degree of disorder and all other letters have their usual meanings. The corresponding energy band diagram is shown in Figure 3.8. This figure indicates that the variation of the

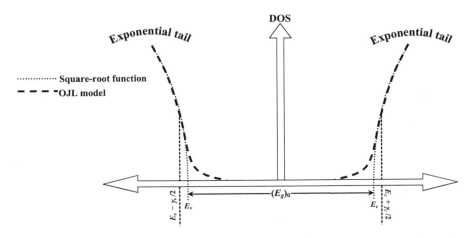

FIGURE 3.8 Energy band diagram showing the exponential and a square-root functional dependence of DOS.

DOS of a perfect crystal can be expressed by a parabolic curve and the corresponding electronic band gap is given by the expression:

$$(E_g)_0 = E_c - E_v \tag{3.8}$$

In Figure 3.8, the thin dotted and thick dashed curves represent band structure for the crystal and amorphous materials, respectively.

Finally, we express the absorption coefficient α in terms of the optical band gap E_g by using the following relation [42]:

$$\alpha h\nu = A(h\nu - E_g)^\eta \tag{3.9}$$

In Equation 3.9, the constant A signifies the features of the bands. This relation also indicates that there is a parabolic rise in the absorption coefficient α with the frequency above the elementary edge.

Here η is transition-related parameter. Its values are 0.5 (i.e., $\eta = 1/2$) and 1.5 (i.e., $\eta = 3/2$) for direct allowed and forbidden transitions, respectively. Further, its values are 2 and 3 for indirect allowed and forbidden transitions, respectively. In most of the cases, an extrapolation of a linear function of $(\alpha h\nu)^{1/2}$ against $h\nu$ [43] is used to define the energy gap of amorphous semiconductors. Thus, a positive relevance of this relation is that we can measure the optical band gap of the amorphous semiconductors by linear fitting the data of $(\alpha h\nu)^{1/2}$ and $h\nu$ [43]. This approach has now become the standard method for the determination of E_g. The enhanced accuracy and the convenient approach of this method make them more reliable as compared to the former old conductivity method. This method can disclose various additional characteristics of the band structure that cannot be identified by using the conductivity method.

3.5 APPLICATIONS OF AMORPHOUS SEMICONDUCTORS

The attention towards the disordered solids has been deeply motivated by the huge number of applications, which depends upon the features that are directly related to the existence of disorderness and defects in the amorphous materials. In this section, we will discuss in brief about some significant applications of amorphous materials. Such applications have been illustrated in Figure 3.9.

Chalcogenide glasses are one of the potential members of family of amorphous semiconductors that have extensive variety of applications in both optics and electronics as well as in optoelectronics. Amorphous thin films of selenium consist of unique features of high resistivity and photoconductivity. These properties make amorphous selenium (a-Se) useful as electro-photographic materials for the respective electronic charging and discharging in the dark and photo-illuminated states. This process is well known as xerography [44,45]. Chalcogenide glasses are also used as phase change memories for recording and erasing in compact disc (CD) and digital versatile disk (DVD) [46].

The solar cells of hydrogenated amorphous silicon (a-Si:H) are the additional examples of amorphous semiconductors. Amorphous silicon (a-Si) is not appropriate to be employed for designing the solar cells because of small lifetime of the charge carriers in a-Si due to the presence of a large number of dangling bonds. In a-Si:H, most of the dangling bonds are saturated by hydrogen; therefore, a-Si:H possesses excellent properties (enough carrier lifetime and sufficient absorption of sunlight) for thin-film solar cells. Such solar cell consists of a transparent, conductive front layer,

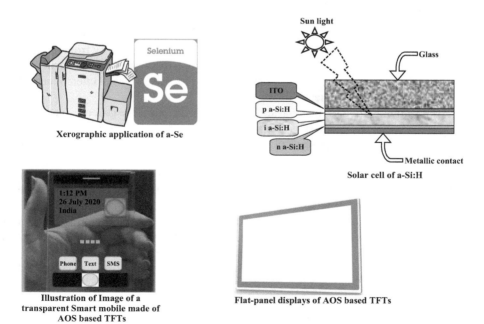

FIGURE 3.9 Schematic for some remarkable applications of amorphous semiconductors.

followed by a p-doped a-SiC:H-layer, the absorber layer of intrinsic a-Si:H, and an n-doped a-Si:H-layer, forming a p-i-n-structure. The two p-doped and n-doped layers create an electric field used to split the charge carriers. The upper, p-doped layer should absorb only slight amount of light because the carriers created in this layer recombine swiftly. This condition can be attained by alloying the silicon with carbon. Such cells are made by using thin-film transistors (TFTs), and they have drawn a great attention of technologists as an alternative supply of energy [47]. These TFTs are also found useful in making flat-panel displays of TVs and monitors [48,49].

Amorphous oxide semiconductors (AOS) have drawn much attention owing to their unique electron transport properties. Like a-Si:H, the TFTs made by AOS are also used widely for flat-panel displays. When we compare the TFTs made by AOS and a-Si:H, we find that AOS-based TFTs are superior because of enhanced pursuance [50]. AOS based TFTs reveal adequate functional characteristics even if we fabricate them at ambient temperature. They have broad dealing out temperature window, large electron mobilities, and low operation voltage [51]. In the near future, the potential and promising applications of AOS are found in the form of fast and sharp-edged flat-panel displays as compared to that are available at present made of a-Si [52].

3.6 PRESENT STATUS

3.6.1 Our Understanding in This Area

The current theoretical and experimental work has improved our perceptive about the various properties of the amorphous semiconductors by clarifying greatly the various unusual characteristics. It is now obvious that the intrinsic defects present in these materials control their electrical properties which are either mandatory from thermodynamic point of view or they can be provoked through the strains created during the deposition process. The chemistry of the ingredient atoms is very well interconnected to the character of these defects.

After studying these semiconductors for about thirty years, the general conclusion in this area is that these materials are important for many new applications instead of conventional applications of semiconductors. Though many attempts have been made to use them for the same applications as in crystalline semiconductors, the devices made from these materials are not found to have the same quality as in case of crystalline semiconductor devices. Only in case of amorphous silicon solar cells, these semiconductors have been found more useful.

This also is not because of better efficiency than crystalline silicon solar cells but because of low-cost involvement in making these cells due to simple technology along with automation in fabrication. Since low-cost conversion is the basic requirement in solar energy application, these materials are considered better as compared to single crystal materials for this particular application.

The application of amorphous semiconductors in conventional devices is restricted due to high density of intrinsic defects, which are automatically created due to rapid cooling of the materials. Efforts are on in preparing materials having low density of defects so that these materials could also be used for all applications where crystalline

Amorphous Semiconductors

semiconductors are being used. In case of amorphous silicon, such defects are reduced by the presence of "H" during preparation or by post hydrogenation. However, similar technique does not work in case of other materials.

The theoretical understanding of these materials is not sufficient to explain many experimentally observed facts as mentioned earlier in this chapter. The reason for this crisis lies in the fact that the density of defects and its distribution with energy is not known exactly. It is therefore very difficult to derive a theoretical formula for any quantity, which depends strongly on the density of defects.

In spite of the above difficulties, some important conclusions have been established regarding these materials:

- These semiconductors have electronic conduction except in a few cases where ionic contribution is quite significant.
- They possess the short-range ordering that is analogous to their crystalline counterpart.

In contrast to earlier belief, these semiconductors could be doped n or p types if the density of defects is low. In Se-type materials, structural defects are paired which explains many experimental findings. The origin of switching phenomenon is well understood, and memory materials are being successfully used in computer memories. Observation of AC conductivity and dielectric loss is well understood in terms of hopping processes in defect states in the band gap of these materials.

3.6.2 Problems for Further Research in This Area

In spite of serious efforts made by various workers in the area of amorphous semiconductors, the following problems are yet to be solved so that these materials could be exploited for device applications in more rigorous way. The problem of reducing structural defects has not yet been solved except in case amorphous silicon. More efforts should be made in this direction. More work is required to understand the light-induced effects observed in these materials. Experimental results are highly intriguing, and hence, more experimental work is required to arrive at a conclusion.

The problem of efficient doping and impurity effects should also be taken further to get more suitable materials for semiconductor devices. More work is required to understand the optical behavior of chalcogenide glasses, which have recently shown photo-induced double refraction phenomenon. Optoelectronic applications must be exploited further for their better use in this area. Amorphous to crystalline transition and the effect of light on crystallization should be studied more thoroughly for getting better materials for optical memories, etc.

The knowledge of exact electronic structure of a particular material requires more theoretical and experimental efforts, which should be made more seriously. Unless the electronic structure is known exactly, no further theoretical work can be done in this area. Search for new, interesting, and potential materials and compositions must continue with a hope to get better and better materials for applications. This is how this field has grown in the past also.

ACKNOWLEDGMENTS

NM is grateful to the Council of Scientific and Industrial Research, New Delhi, India, for providing financial assistance under the Research Scheme for Physical Sciences (Scheme no. 03(1453)/19/EMR-II).

REFERENCES

1. B. T. Kolomiets, and N. A. Goryunova, *Zhurn. Techn. Fiz.* 25, 984 (1955).
2. S. R. Ovshinsky, *Phys. Rev. Lett.* 21, 1450 (1968).
3. M. H. Cohen, H. Fritzsche, and S. R. Ovshinsky, *Phys. Lett.* 22, 1065 (1969).
4. J. P. De Neufville, S. C. Moss, and S. R. Ovshinsky, *J. Non-Cryst. Sol.* 13, 191 (1973/1974).
5. S. Kugler, K. Shimakawa, *Amorphous Semiconductors.* Cambridge University Press: Cambridge (2015).
6. K. Tanaka, A. Maruyama, T. Shimada, H. Okamoto, *Amorphous Silicon.* John Wiley & Sons: Chichester, (1999).
7. L. C. Chen, and F. Spaepen, *Nature* 336, 366 (1988).
8. D. Adler, *Amorphous Semiconductors.* CRC Press: Cleveland, OH (1971).
9. B. T. Kolomiets, *Phys. Stat. Sol.* 7, 359 (1964).
10. N. F. Mott, E. A. Davis, and R. A. Street, *Philos. Mag.* 32, 961 (1975).
11. R. A. Street, N. F. Mott, *Phys. Rev. Lett.* 35, 1293 (1975).
12. S. C. Agrawal, *Phys. Rev. B* 7, 685 (1973).
13. H. Folich, *Pro. Roy. Soc. A* 188, 521 (1947).
14. L M. Lifshitz, *Adv. Phys.* 13, 483 (1964).
15. N. F. Mott, and E. A. Davis, *Electronic Processes in Non-Crystalline Materials.* Clarendon Press: Oxford, 428 (1979).
16. M. H. Cohen, H. Fritzsche, and S. R. Ovshinsky, *Phys. Rev. Lett.* 22, 1065 (1969).
17. E. A. Davis, and N. F. Mott, *Philos. Mag.* 22, 903 (1970).
18. N. F. Mott, *Phil. Mag.* 19, 835 (1969).
19. N. F. Mott, *Conduction in Non-Crystalline Materials*, 2nd ed. Clarendon Press: Oxford (1992).
20. P. W. Anderson, *Phys. Rev. Lett.* 34, 953 (1975).
21. K. Morigaki, S. Kugler, and K. Shimakawa, *Preparation Techniques, in Amorphous Semiconductors: Structural, Optical, and Electronic Properties.* John Wiley & Sons, Ltd: Chichester. Chapter 2, (2017).
22. N. Mehta, *J. Sci. Indust. Res.* 65, 777 (2006).
23. J. A. Rowlands, and S. O. Kasap, *Phys. Today* 50, 24 (1997).
24. A. E. Delahoy, and S. Guo, Transparent conducting oxides for photovoltaics. In *Handbook of Photovoltaic Science and Engineering*, 2nd ed., A. Luque and S. Hegedus, eds. John Wiley & Sons: Chichester, 716–796 (2010).
25. C. C. Koch, O. B. Calvin, C. G. Macklamey, and J. O. Scarbrough, *Appl. Phys. Lett.* 43, 1017 (1983).
26. R. B. Schwarz, and C. C. Koch, *Appl. Phys. Lett.* 49 146 (1986).
27. J. L. Rojas, M. Dominguez, P. Villares, and R. Jimenez-Garay, *Mater. Chem. Phys.* 45, 75 (1996).
28. M. Roilos, *Philos. Mag. Part B* 38, 477 (1978).
29. W. Beyer, J. Stuke, *J. Non-Cryst. Sol.* 8–10, 321 (1972).
30. S. R. Elliott, *Adv. Phys.* 36, 135 (1987).
31. N. Mehta, A. Dwivedi, R. Arora, S. Kumar, and A. Kumar, *Bull. Mater. Sci.* 28, 579 (2005).

Amorphous Semiconductors

32. I. Solomon, R. Benferhat, and H. T. Quoc, *Phys. Rev. B* 30, 3422 (1984).
33. N. Chandel, N. Mehta, and A. Kumar, *Vacuum* 86, 480 (2011).
34. N. Chandel, N. Mehta, A. Kumar, *J. Electron. Mater.* 44, 2585 (2015).
35. L. S. Miller, A. J. Walder, P. Linsell, and A. Blundell, *Thin Solid Films* 165, 11 (1985).
36. J. P. Borgofro, B. Lazarides, and E. Pelletier, *Appl. Optics* 21, 4020 (1982).
37. S. V. Babu, M. David, and R. C. Pate, *Appl. Optics* 30, 839 (1991).
38. S. Hasegawa, S. Yazaki, and T. Shimizu, *Sol. Stat. Commun.* 26, 407 (1978).
39. F. J. Blatt, *Physics of Electron Conduction in Solids.* McGraw-Hill: New York (1968).
40. K. L. Chopra, *Thin Film Phenomena.* McGraw-Hill: New York (1969).
41. S. K. O'Leary, S. R. Johnson, and P. K. Lim, *J. Appl. Phys.* 82, 333, (1998).
42. J. I. Pankove, *Optical Processes in Semiconductors.* Prentice-Hall, Englewood Cliffs, NJ, p. 34 (1971).
43. J. Tauc, Optical properties of amorphous semiconductors. In *Amorphous and Liquid Semiconductors.* Springer: Boston, MA, 1974, pp. 159–220.
44. D. M. Pai, and B.E. Springett, *Rev. Mod. Phys.* 65, 163 (1993).
45. P. Selenyi, *Z. Tech. Phys.* 12, 607 (1935).
46. N. Mehta, *Rev. Adv. Sci. Eng.* 4, 173 (2015).
47. D. E. Carlson, and C. R. Wronski, *Appl. Phys. Lett.* 28, 671 (1976).
48. M. J. Powell, *MRS Symp. Proc.*, 33, 259 (1984).
49. W. E. Spear, and P.G. LeComber, *Semicond. Semimetals* 21D, 89 (1984).
50. T. Kamiya, and H. Hosono, *NPG Asia Mater.* 2, 15 (2010).
51. K. Nomura et al., *Nature* 432, 488 (2004).
52. J. F. Wager, and R. Hoffman, *IEEE Spectrum* 48, 42 (2011).

4 Promise of Self-lubricating Aluminum-Based Composite Material

Neeraj Kumar Bhoi, Harpreet Singh, and Saurabh Pratap
Indian Institute of Information Technology, Design and Manufacturing

CONTENTS

4.1 Introduction 65
4.2 Historical Background and Need for the Development of Al-SLMMCs 67
4.3 Wear Mechanism 69
    4.3.1 Wear Measurement Techniques 71
    4.3.2 Erosive Wear 72
    4.3.3 Reciprocating Wear 72
    4.3.4 Pin on Disc 73
    4.3.5 High- and Low-Stress Wear Test 73
4.4 Fabrication Methods for Al-SLMMCs 73
4.5 Reinforcement for Self-lubricating Behavior 74
4.6 Correlation between Mechanical and Tribological Aspect of Al-SLMMCs 76
4.7 Conclusions 81
References 81

4.1 INTRODUCTION

In order to maintain material properties to be stable, materials are subjected to different functional environment from largest to nano length scale. The process followed for the stable phase development over the substrate material always been a challenging task by the consideration of fundamental material configuration and chemical composition imposed upon the surface. The presence of oxides, carbides and several inter-metallic compounds on the surface largely enhances the hardness, fracture toughness and wear resistance of the material [1]. The enthusiasm to play with the surface texture of the material by the additions of different layers over the surface material always had been a quest for the research community. Crafting for the improved performance in surface chemistry is a key concern in the mechanical and tribological industry. High wear and corrosion-resistant materials draw larger interest in the various sectors such as automobile, aerospace, naval and spacecraft industries [2]. Surface modification is a major breakthrough in the numerous industrial civilizations with higher productivity and improved functional performance [3]. However, the requirement for higher specific energy, larger processing time and numerous process variables needs to be controlled for better functional application. Composite material tailors and combines the best properties of the matrix and supports element by providing (i) ductility and toughness of the matrix and (ii) strength and modulus of the reinforcement element for different structural and functional applications [4,5].

In the applied area of the tribological component, most of the materials struggle between the different phenomenon of friction and wear process. However, this can be greatly minimized by the virtue of the lubricating medium between the mating surfaces. The lubricant material provides the shearing between the mating part that, in turn, reduces the friction coefficient and provides better sliding actions, which results in increased life to the component. The shear strength value of the lubricating layer is less than the mating part. Therefore, the lower value of the strength reduces the friction value. The major challenges arise when the material has to create an extreme work environment such as a vacuum, extremely low-temperature zone and extreme contact pressure area [6]. This situation can be eliminated by the use of self-lubricating material (SLM), which helps in reducing the friction between the contact area. The function of the SLM is similar to that of applied lubricating medium between the counterpart. SLM provides the low shear strength between the surface which in turn minimal friction between the mating parts. To enhance the better and more lubricity between the surfaces, the synergy of lubricating medium and SLM can be utilized. This process can have a better influence on the wear performance of the material.

Composite material has been used since the Stone Age by the combination of two or more elements to meet the customer requirement. The use of composite material can be found in different applications from day–to-day life and future novel product applications. The composite material is largely employed in several tribological applications where the control of friction and wear is of major concern [7]. Aluminum-based self-lubricating composite material (Al-SLMMCs) can be found in the different tribological components, as well as the demands in various wear-resistance behavior are of primal considerations. The Al-SLMMCs use the hard ceramics, transition metal oxides, rare earth oxides (as reinforcement), which are typically used to modify the functional property of the material as well as for the reduction in the wear rate of the material. Al-SLMMCs take the advantages of the hard phase of the reinforcement and the lubricating behavior of the material to reduce the shear strength between the mating part. The reinforcement provides the thin tribo-layer or metal matrix layer (MML) between the surfaces for the reduction in the wear behavior of the material.

Al-Based Self-lubricating Composite Material

4.2 HISTORICAL BACKGROUND AND NEED FOR THE DEVELOPMENT OF Al-SLMMCs

The use of lubrication started in early 1970 for the reduction in wear in the different mating parts and engine components. In the initial stages of lubrication technology, the crude oil and petroleum refined products were directly implemented over the target area. The demands for the high-temperature machining and hybrid manufacturing processes lead to the development of different soft and hard coatings over the surface for reconfiguration as per the requirement. The use of micro-electromechanical system (i.e. different nano components) in the numerous electronics sectors requires the need of solid lubricating material which provides the uniform and better lubricity during functioning. The use of solid lubrication started in the early 1990s for different machine components. The automobile, aerospace and marine applications are the highly demanding areas for the use of solid lubrication material technology. Table 4.1 represents the different stages and uses of lubrication technology in the various sectors of day-to-day life. A comparative assessment of the historical background of the development stages in the self-lubricating composite material,

TABLE 4.1
Different Functional Application in the Historical Development Stages and Self-Lubricating Material System

Year of Development	Type of Material	Applications
Early 1970	Crude oil, petroleum	Directly applied over the required functional surface for different applications, gear box, engine component and industrial parts.
1970–1990	Soft and hard coating	Surface reconfiguration and modifications, Physical vapour deposition (PVD) and Chemical vapour deposition (CVD) process Cutting tool applications, extreme pressure applications, manufacturing operations and processes
	Solid lubricants, organic and in-organic materials	From micro to nano to microstructure components, brake drums, connecting rod, pistons, engine component
1990–2010	Nanostructured solid lubricants, functionally graded composite material, smart functional material	Smart structure, small micro mechanical components, functionally graded implants, space craft, vibrational control
2010–present	Adaptive and smart lubrication, advance self-lubrication system, self-dispersed, improved lubricity and material for aggressive environment, self-decomposed material, eco- and green tribo-materials, concept of cleaner tribology	Advanced supersonic spacecraft, aerospace material, sustainable material applications, functional bio-medical applications, tissue structures

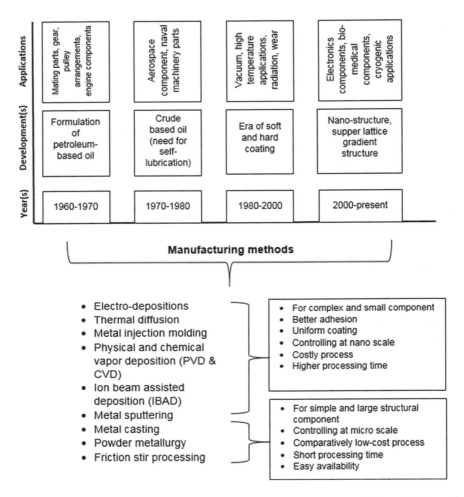

FIGURE 4.1 Development stages in the self-lubricating composite material, application and different manufacturing processes and their advantages and disadvantages.

application and different manufacturing processes and their advantages and disadvantages is given in Figure 4.1. The use of advanced lubricating system, nanostructured coating and superlattice gradient manufacturing development started in the era of 2000s where the manufactures is exposed to different advanced functional applications, from advanced spacecraft to extreme harsh operating conditions in marine applications. The need for the sustainable process drives the researchers and industrial experts for the development of material that can operate under extreme high temperature and pressure application with advanced self-dispersed, self-lubricity and chemically inert material that can be operated under unidentified working environment smoothly.

The graded compositional material for the different tribological applications has been a great interest of all time for the material scientists. The classic example of

Al-Based Self-lubricating Composite Material

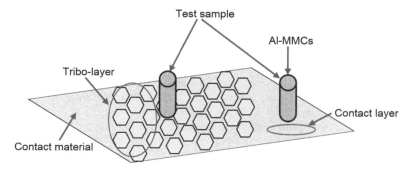

FIGURE 4.2 Generation of tribo-layer during the contact zone of two distinct materials.

SLMMCs is the gray cast iron (GCI) where the hard iron matrix utilizes the dispersed graphite flakes. The dispersion mode of the reinforcement in the matrix defines the structural integrity over the different applications. As discussed in the earlier section, the use of Al-SLMMCs can be found suitable for the different functional applications such as extreme pressure to vacuum mode. For instance, in case of the SLMMCs, the reinforcement transfers the layer of tribo-layer to the contact zone for the reduction in the wear behavior of the material. Figure 4.2 shows the schematic representation of the generation of tribo-layer during the performance of any material under contact with the hard surface. During the lubricated environment or material having lubricating effect, a thin tribo-layer is formed in the contact zone, which prevents the material against failure due to wear action.

The concept of the generation of the material which formed a tribo-layer during the contact mode without damaging the inherent nature of the material is a challenging task to look upon. Smooth transition in the micro-structure for the development of material under a special class of Al-SLMMCs has been a big goal for the researchers around the globe that will accommodate the specified work under different extreme environment. This chapter will drive the researchers and industrial application for the development of the material that will serve the improved class of wear resistance for different applications.

4.3 WEAR MECHANISM

The wear of any material can be defined as the loss of material due to contact in two body pair. The wear behavior of the material can be classified as per the application and nature of the wear. The different types of wear and their nature along with the area where the particular phenomenon is observed are tabulated. Table 4.2 narrates the different types of wear observed.

The wear behavior of any material is affected by several parameters from the selection of raw material to the actual operating environment of the material. The main affecting parameter during the performance of the Al composite can be given as:

1. Matrix and reinforcement factor.
2. Physical and mechanical factor.

TABLE 4.2
Different Types of Wear and Their Behavior

Types of Wear	Nature/Behavior	Applicable Area
Erosive wear	Striking of hard ceramic particles to the surface in wet, dry or semi-solid medium	Turbine blade, landing gear, airframe, helicopter blades
Abrasive wear	Presence of hard particles, striking and forced movement of the surface	Sliding part
Sliding wear	Transfer of material from soft to hard surface, material loss, localized banding between the surface	Sliders, gear, cam shaft
Fretting wear	Due to fretting action between the parts in contact	Press fitting
Fatigue wear	Wear generated due to fatigue action in material	Rotor dynamics, bearing parts
Cavitation wear	Erosion causing the presence of vapor phase in the material	Bearing surfaces

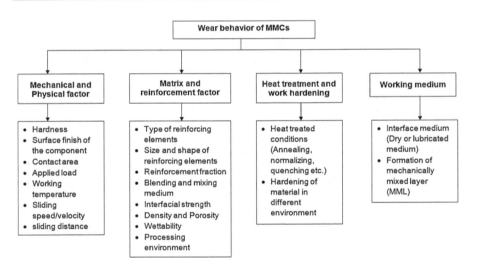

FIGURE 4.3 Factors responsible for the wear in the material.

3. Heat treatment and work hardening.
4. Operating environment and conditions.

The detailed affecting parameters of the stated properties are given in Figure 4.3. Addition of the hard ceramic in the matrix material greatly imparts the wear resistance property of the material. However, after a certain limit, the fracture toughness of the material reduces with the higher content of the ceramic in the matrix, which leads to easy removal of the ceramic from the matrix, eventually higher chances of

Al-Based Self-lubricating Composite Material

the wear of the sample. Thus, increasing the hard-ceramic phase in the material with the compromising fracture toughness causes the dramatic failure of the material.

4.3.1 Wear Measurement Techniques

With the increasing demand in the high-performance material for efficient use in light weight, high-strength aluminum alloy is of greater importance for the designers. To meet the specified standard, the material has to perform in the different surroundings undertaken. The wear behavior of the material is assessed by a number of techniques depending upon the type of application and demands. During the wear measurement of the material, the frequent and most applied way is to find out the weight of the sample before and after the test by means of delicate micro-balance. Figure 4.4 represents

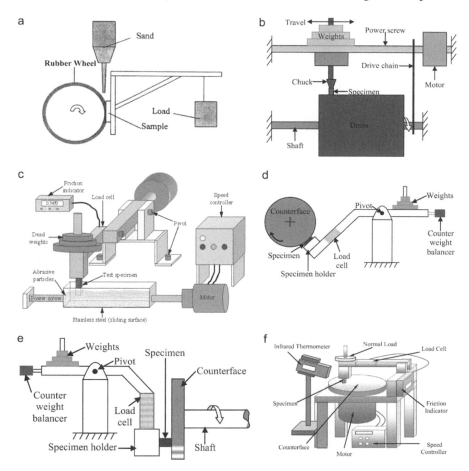

FIGURE 4.4 Schematic illustration of tribo wear test machines. (a) Dry sand rubber wheel. (b) Pin on drum. (c) Linear tribo machine. (d) Block on ring. (e) Pin on disc. (f) Block on disc [7]. (Reprinted from Tribology International, 83, Author(s), Umar Nirmal, Jamil Hashim, M.M.H. Megat Ahmad, 77–104, Copyright 2015, with permission from Elsevier.)

the different types of wear behavior measurement systems for the different composite materials.

The wear rate of the sample can be given by the following expression 4.1:

$$\text{wear rate} = \frac{(\text{weight loss}) \ \text{or} \ (\text{volume loss})}{(\text{sliding distance})} \qquad (4.1)$$

To accommodate the physical property of the material, sometimes the wear number is found suitable for the judgment of the wear-resisting capacity of the material. The wear number of the material can be defined as 4.2:

$$\text{wear number} = \frac{\text{density}}{\text{weight loss}} \qquad (4.2)$$

The higher wear number signifies the better wear-resisting capacity of the material. Sometimes, the same can be assessed by means of calculating the volume loss of the material. The lower volume loss under similar operating environment can be termed as the good wear-resisting material for the adopted set of samples.

$$\text{volume loss} = \frac{\text{weight loss}}{\text{density}} \qquad (4.3)$$

Description of the different ASTM standard for the assessment of the wear behavior of the material is stated as:

4.3.2 Erosive Wear

For the measurement of erosive wear behavior, the solid hard particles are impregnated to the test sample by some mechanical action or by belt pulley arrangements. For the measurement ASTM, G76 standard is used for the different materials. The main affecting process parameters during the test are impingement angle, applied pressure range, standoff distance and size of the erodent particles. The erosive behavior test system is given in Figure 4.4a.

4.3.3 Reciprocating Wear

For reciprocating wear behavior, ASTM G99-95a is used for the preparation of the test sample. Similar to pin on disc test, the counterpart consists of a harder phase than sample material, and hardness of the plate is taken above the HRC 55. During this test condition, the sample or counterpart is to be reciprocated for the determination of material behavior under reciprocating mode. The stroke frequency is kept in the range of 30–60 Hz with the reciprocating speed that varies from minimum 0.1 m/s to maximum of up to 10 m/s. A schematic representation of the test setup is shown in Figure 4.4c.

Al-Based Self-lubricating Composite Material 73

4.3.4 Pin on Disc

The pin disc apparatus or method is widely employed for the quantification of the wear rate and friction co-efficient of the material. Under this method, the test sample is placed in contact with the hard surface (*typically the Rockwell hardness-C (HRC) of the counterpart is equal to or more than 55*). The pin on disc test is carried out under dry or lubricated environment and in the normal or elevated temperature range. The friction co-efficient value can be found at any point in the test by the calculation of the normal applied load. For this test, the ASTM G99-95a is applied with the sample size ranges between 3 and 25 mm in diameter and height in the range of 5–25 mm. A typical arrangement of the pin on disc test apparatus is given in Figure 4.4e.

4.3.5 High- and Low-Stress Wear Test

The test is similar to the pin on disc test except the high-stress wear test uses the slurry of coarse alumina particles (i.e. 500–600 micron) as the abrasive medium between the sample and counterpart, and low-stress wear test uses the dry quartz sand particle (i.e. 200–300 micron) between the sample and counterpart. ASTM B611 and ASTMG65 are utilized for high- and low-stress wear tests, respectively. A rotating steel counterpart is applied during the high-stress wear test while the rubber wheel is used during low-stress wear behavior analysis. In the stated test, wear number and volume loss are assessed for the quantification of the wear resistance of the test sample.

For the enhancement in the wear behavior, hardness plays a key role in the overall performance. As per the Archard law, the wear co-efficient of material is governed by Equation 4.4.

$$V = k \frac{WL}{3H} \tag{4.4}$$

Where, V is volumetric loss of the work material, k is the wear co-efficient, W is applied load, L is the total sliding distance, and H is the hardness of the work material.

4.4 FABRICATION METHODS FOR AL-SLMMCs

Every mechanical system needs some lubricating property for the proper functioning under the different operating regimes. The simple and easy way to create a lubricating medium between the contact zone of the mating part is by the application of liquid lubrication.. However, in case of small mechanical components, internal IC engine parts and delicate instruments in the electronics industry, it is not possible to create a lubricating environment. Surface coating and cladding approach are another way to generate the lubricating zone on the functional surface [8]. However, the coating is limited to a certain height in the component, and life of them is limited. This type of problem can be reconfigured by means of generating a material that can have the ability to provide the lubricating medium during operation. This type of material is kept under the category of self-lubricating material [9,10]. The self-lubricating type material can be defined as the material that can able to provide lubricating effect under different speeds and load without damaging (i.e. particle abrasion, cold welding, adhesion or erosion). The material damaging observed in body pair when they are in contact or rub against each other under dry or any specified lubricating medium With the absence of lubricating medium between the contact and mating parts material damaging is observed. The development of self-lubricating type composite material is of a great challenge due to lack of correlation between the physical, chemical, mechanical and tribological properties of the material. The friction co-efficient and wear resistance are a major and prime consideration for the development of self-lubricating type material. However, other associated properties must be taken into consideration for the wide applicability in the product. Self-lubricating composite material can be prepared by means of applying liquid metallurgy or solid-state metallurgy. The simple and easy way for the generation of SLM is the adoption of a powder metallurgy process. The consolidation behavior during the powder metallurgy process has been commercialized for many decades for different tribological applications [11].

For the development of Al-SLMMCs, two main processing techniques are applied: liquid-state processing (casting) and solid-state processing (powder metallurgy). During liquid-state processing, the matrix material (*here aluminum and its alloys*) is brought into the molten state by means of one or different heating mode. In the molten state, the reinforcing elements are added and mixed by stirring action or by centrifugal action. The stirring action is done for the prevention of reinforcement agglomeration in the matrix material. The mixed material is finally poured to the different mold as per the requirement and use. During powder metallurgy, the initial matrix and reinforcing elements are mixed by means of blending or mixing unit. The blending process improves the diffusion of the reinforcement with the matrix material. The blended powder is further subjected to different compaction system and sintering mode for the final product development. The sintering process improves the particles diffusion and grain growth eventually leading to the uniform densified product with improved material properties [5,12]. The affecting process parameter during casting and powder metallurgy process is mentioned in Figure 4.5.

4.5 REINFORCEMENT FOR SELF-LUBRICATING BEHAVIOR

The basic aim while fabricating the self-lubricating type composite material is to care about the selection of the manufacturing process and the volume fraction of the reinforcement. The key behind the success of any self-lubricating material lies in the property and distribution of the reinforcement in the material. The improper distribution of the particle inside the matrix has a detrimental effect on output and can cause a number of disasters in running life. The higher volume fraction of the reinforcement may have better lubricity and wear resistance compared to matrix material. However, with the higher amount of volume percentage in the matrix has a harmful effect on the bulk metallic property. Hence, the optimum level of reinforcement in the matrix defines the performance of the product and for the proper harmonizing of the different properties. Most of the researchers used the addition of hard ceramics upto 20 wt.% in the matrix as the higher hard ceramic phase in material largely enhances

Al-Based Self-lubricating Composite Material

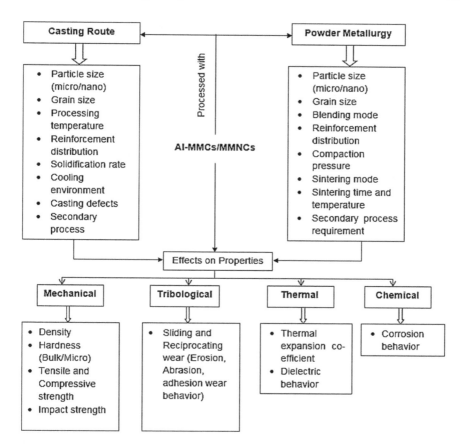

FIGURE 4.5 Process variable during casting and powder metallurgy techniques.

the wear resistance [13]. However, after 10 wt.% of the reinforcement, the mechanical behavior (i.e. strength and ductility) of the developed material tends to decrease due to increasing behavior in the material brittleness. In the field of Al-MMCs, numerous reinforcing elements were utilized for the enhancement of the tribological response of the material. A beautiful depiction of the types of reinforcement utilized for the self-lubricating type composite material is given in Figure 4.6 [14].

In Al-SLMMCs, carbon-based reinforcement is largely employed for the improvement of mechanical as well as the tribological response of the material. Carbon nanotube (CNT)-based reinforcement in the form of single-walled carbon nanotube (SWCNT) and multi-walled carbon nanotube (MWCNT) greatly imparts the material behavior with slight addition in the matrix material [15]. For Al-SLMMCs, oxides, carbides, borides and nitrides are effectively utilized for the enhancement of the wear response of the material. However, research in the direction of halides, metallic and in-organic material is not being observed carefully. Earlier research in the past suggested that a very small amount (i.e. say in the fraction of 0.01–1 wt.%) of the carbon-based material reinforcement improves the wear resistance significantly [16].

76 Functional and Smart Materials

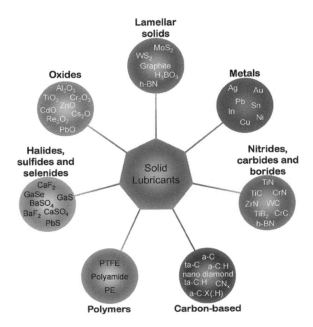

FIGURE 4.6 Examples of the different solid lubricants [14]. (Reprinted from Tribology International, 120, Kaline Pagnan Furlan, José Daniel Biasoli de Mello, Aloisio Nelmo Klein, Self-lubricating composites containing MoS2: A review, 280–298, Copyright (2018), with permission from Elsevier.)

The concept of encapsulating the reinforcement with the matrix material improved the material response significantly to a greater extent. The capsulizing lubricating material over hard ceramic content improves the lubricity over the surface with the simultaneously maintaining the mechanical properties of the material.

4.6 CORRELATION BETWEEN MECHANICAL AND TRIBOLOGICAL ASPECT OF Al-SLMMCs

The major challenging test during the processing of Al-SLMMCs is a uniform distribution of the reinforcement in the matrix material. Without altering the mechanical property of the aluminum, the wear resistance behavior improvement is a tough and challenging task to control. Numerous reinforcing ways in the form of lamellar solids, ceramics, in-organic compounds, a carbon-based material, hard transition metallic material are utilized in the past. The use of lamellar solids improves the wear resistance significantly with the conciliation of the strength and ductility [17]. The addition of the graphite and dual ceramic phase (i.e. hard ceramic and graphite) in the matrix material maximize the material performance with a significant reduction in the wear. The co-efficient of friction drastically reduces with the addition of graphite in the reinforcement [18]. The statement is depicted in Figure 4.7 where aluminum oxide and graphite are added for the fabrication of composite material. Tables 4.3 and 4.4 represent the various compositions and mechanical properties of the material. During

Al-Based Self-lubricating Composite Material

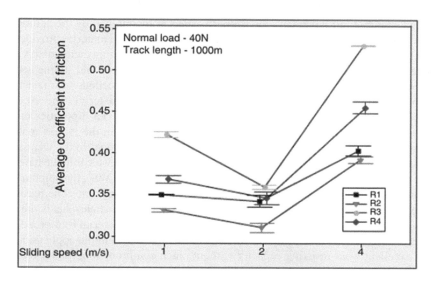

FIGURE 4.7 Representation of average coefficient of friction of composites for high category [18]. (Open access article under creative common license; require no permission.)

TABLE 4.3
Composition of Reinforcement [18]

Reinforcement	Composition
R1	20wt.% Al_2O_3/20wt.%Si_3N_4/55wt.% Al/5wt.% graphite
R2	10wt% Al_2O_3/30wt%Si_3N_4/55wt% Al/5wt% graphite
R3	45wt.% Al_2O_3/50wt.%Al/5 wt.% graphite
R4	95wt.% Al_2O_3/5wt.% graphite

TABLE 4.4
Mechanical Properties of the Composites [18]

Composites	Ultimate Tensile Strength (MPa)	Ultimate Compressive Strength (MPa)	Vickers Hardness (HV)	Ductility (%)
R1	131.84	292.6	62.2	5
R2	153.78	278.49	61.5	8.3
R3	96.16	361.74	66.3	2.5
R4	44.14	298.2	59.9	1.25

the dual ceramic reinforcement, the strength aspect is maintained by the addition of a hard-ceramic phase, and the graphite particles present in the material provides the shearing action between the contact zone. The graphite creates tiny metal matrix layer between the rubbing medium which provides a smooth transition and sliding action.

The use of in-organic material in the aluminum exhibits excellent wear resistance without much compromising the mechanical aspect of the material. The excellent use of in-organic reinforcement and their effect on the material properties can be seen stunningly in Figure 4.8. The use of WS_2 and $2H-WS_2$ in the matrix material slightly decreases the elastic modulus, but excellent improvement in the wear resistance can be observed. The tensile modulus reduces up to 30.13% while the hardness value reported being increased as 24.44% [8]. With the use of WS^2, the sulphur containing anti-wear spices create a tribo-film layer in the contact zone. The reduction in the friction coefficient with the use of WS_2 and $2H-WS_2$ particles states the controlling capabilities of the composite material. The organic material reinforced with composite material is much suitable for the high-temperature engine application due to its excellent wear-resisting capacity without much compromising the mechanical aspect of the material.

The important factors affecting the tribological response of the aluminum metal matrix composites are the type of reinforcing particles, sizes, a fraction of the particles and most importantly the distribution in the matrix material. The distribution and size of the particles define the outcomes and performance of the material under critical operating conditions [19]. A typical comparison between the premixed and pre-alloyed aluminum powder was made with TiB_2 as the reinforcement. It was observed that the higher volume fraction of the TiB_2 reduces the average friction co-efficient value. The higher applied load in the sample material increases the amount of wear rate. In the higher reinforcement and increased loading conditions, the contact area between the mating parts is increased, eventually the transfer layer is strongly compacted which results in generation of higher MML layer. [20]. Figure 4.9 shows the wear of $LM4-TiB_2$ reinforced as a function of reinforcement percentage for different rotational speed. The SEM micrographs of the worn-out surface are shown in Figure 4.10. At smaller reinforcement percentage, the more

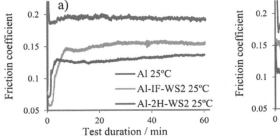 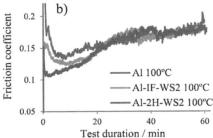

FIGURE 4.8 Average friction coefficient of the stroke length during the tests performed at 25°C (a) and 100°C (b) [9]. (Open access article published in Springer Nature require no permission.)

Al-Based Self-lubricating Composite Material

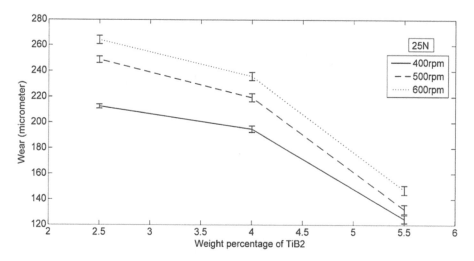

FIGURE 4.9 Wear as a function of weight percentage of TiB$_2$ for varying rotational speed [21]. (Reprinted by permission from Silicon: Springer Nature, Tribological Characterization of Stir-cast Aluminium-TiB² Metal Matrix Composites by Suswagata Poria, Prasanta Sahoo, Goutam Sutradhar, Copyright 2016.)

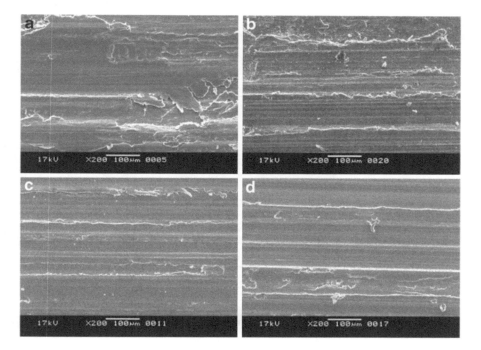

FIGURE 4.10 SEM images of worn surfaces of Al-TiB$_2$ composites (a) 1%, (b) 2.5%, (c) 4% and (d) 5.5% [21]. (Reprinted by permission from Silicon: Springer Nature, Tribological Characterization of Stir-cast Aluminium-TiB² Metal Matrix Composites by Suswagata Poria, Prasanta Sahoo, Goutam Sutradhar, Copyright 2016)

plowing of the surface is observed for the different composite materials [21]. In the similar context, the wear behavior of the aerospace alloy (AA5052) and in situ TiC as reinforcing element was fabricated by stir casting method for superior mechanical and tribological behavior. It was observed that material containing 9% TiC exhibits improved performance in terms of strength and ductility. The hardness and ultimate tensile strength value are improved by 32% and 78%, respectively. Further, the sliding wear behavior of the material shows decreased friction co-efficient and improved resisting capacity for the Al/TiC composite material [22]. In case of hard ceramic content, the restriction of material flow from the composite surface prevents wear and damage. These phenomena can be explained by the formation of solid solution during the reinforcement, and the interfacial bonding plays a vital role in the improvement of mechanical and tribological properties of the material [4,23–25].

The dry sliding behavior study of the 20vol.% Al_2O_3 in aluminum exhibits better wear resistance compared to the monolithic phase of the material. It was observed that higher content of reinforcement and higher sliding distance can reduce the wear rate and friction co-efficient value [25]. A similar observation made by a group of researchers suggests that the use of dual-phase reinforcement (i.e. hybrid composite) improves the material performance to a great extent. A beautiful representation of the effect of hybridization can be seen from Figure 4.11. Under a similar operating environment, the use of small fraction of Al_2O_3 and SiC in the monolithic phase improves the wear resistance to a larger scale compared to higher content of reinforcement [24].

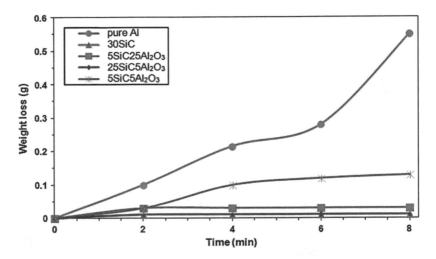

FIGURE 4.11 Wear behaviors of aluminum and hybrid MMCs at load 5 N [23]. (Reprinted by permission from Acta Metallurgica Sinica (English Letters): Springer Nature, Tribological Characterization of Hybrid Metal Matrix Composites Processed by Powder Metallurgy by M. Megahed, M. A. Attia, M. Abdelhameed et al, Copyright 2017.)

4.7 CONCLUSIONS

Development of cost-effective method for surface and structure modification without compromising the functionality of the bulk material is the major challenging work in material processing. Surface modification with coating is the possible remedies for the generation of lubricating environment over substrate material. However, the functionality in coating is limited to certain products and specific applications with certain thickness. The possible remedies for overcoming the problem generated during the coating/cladding are the development of material that provides the better wear-resisting capacity as well as the lubricating effect at operating conditions. The SLMMCs will count and accommodate the overall microstructure and property at the bulk and small scale. As per the description stated in this chapter, the following major points can be given:

1. The use of hybridization in the aluminum matrix holds strong interfacial bonding, good structural integrity and better wear resistance properties. However, more extensive research is required in this direction to utilize the different phases of the reinforcement for the development of self-lubricating type material.
2. The mechanism exists during the reinforcement needs more exploratory research for the development of better lubrication in the operating environment. Use of in-organic based materials for wear resistance applications must be explored for the insurance of better performance under different complex working phases.
3. Online monitoring and controlling of process parameter in the operating region must be utilized effectively for the better service life and product reliability.

REFERENCES

1. X. Ma, Y.F. Zhao, W.J. Tian, Z. Qian, H.W. Chen, Y.Y. Wu, X.F. Liu, A novel Al matrix composite reinforced by nano-AlN p network, *Sci. Rep.* 6 (2016) 1–8. doi: 10.1038/srep34919.
2. M. Tabandeh-khorshid, E. Omrani, P.L. Menezes, Tribological performance of self-lubricating aluminum matrix nanocomposites : Role of graphene nanoplatelets, *Eng. Sci. Technol. an Int. J.* 19 (2016) 463–469. doi: 10.1016/j.jestch.2015.09.005.
3. N. Saini, D.K. Dwivedi, P.K. Jain, H. Singh, Surface modification of cast Al-17%Si alloys using friction stir processing, *Procedia Eng.* 100 (2015) 1522–1531.
4. N.K. Bhoi, H. Singh, S. Pratap, Developments in the aluminum metal matrix composites reinforced by micro / nano particles – A review, *J. Compos. Mater.* (2019) 1–21. doi: 10.1177/0021998319865307.
5. H. Singh, P.K. Jain, N. Bhoi, S. Pratap, Experimental study pertaining to microwave sintering (MWS) of Al-metal matrix composite - A review, *3rd Int. Conf. Compos. Mater. Mater. Eng.* 928 (2018) 150–155. doi: 10.4028/www.scientific.net/MSF.928.150.
6. D.G. Backman, J.C. Williams, Advanced materials for aircraft engine applications, *Science* 255 (1992) 1082–1087. doi: 10.1126/science.255.5048.1082.
7. A.M.M.H.M. Nirmal Umar, Hashim Jamil, A review on tribological performance of natural fibre polymeric composites, *Tribol. Int.* 83 (2015) 77–104.
8. H. Singh, P.K. Jain, The current issues and challenges of product recovery of mating surfaces, B. Katalinic (Ed.), Vienna: DAAAM International Scientific Book, Chapter 29, 2014, pp. 365–372.

9. V.B. Niste, M. Ratoi, H. Tanaka, F. Xu, Y. Zhu, J. Sugimura, Self-lubricating Al-WS2composites for efficient and greener tribological parts, *Sci. Rep.* 7 (2017) 1–14. doi: 10.1038/s41598-017-15297-6.

10. J. Zhang, Y. Wang, B. Yang, B. Zhou, Effects of Si content on the microstructure and tensile strength of an in situ Al⁻Mg 2 Si composite, *J. Mater. Res.* 14 (1999) 68–74.

11. J. Singh, Fabrication characteristics and tribological behavior of Al/SiC/Gr hybrid aluminum matrix composites: A review, *Friction.* 4 (2016) 191–207. doi: 10.1007/s40544-016-0116-8.

12. N.K. Bhoi, H. Singh, S. Pratap, P.K. Jain, Microwave material processing : A clean, green, and sustainable approach, Kaushik Kumar, Divya Zindani, Paulo Davim (Eds.) in: *Sustainable Engineering Products and Manufacturing Technologies.*, 1st ed., Academic Press Elsevier, 2019: pp. 3–23. doi:10.1016/B978-0-12-816564-5.00001-3.

13. J. Lakshmipathy, B. Kulendran, Reciprocating wear behavior of 7075Al / SiC in comparison with, *Int. J. Refract. Hard Mater.* 46 (2014) 137–144. doi:10.1016/j.ijrmhm.2014.06.007.

14. K.P. Furlan, J.D.B. de Mello, A.N. Klein, Self-lubricating composites containing MoS2: A review, *Tribol. Int.* 120 (2018) 280–298. doi:10.1016/j.triboint.2017.12.033.

15. A.M.K. Esawi, K. Morsi, A. Sayed, A.A. Gawad, P. Borah, Fabrication and properties of dispersed carbon nanotube-aluminum composites, *Mater. Sci. Eng. A.* 508 (2009) 167–173. doi:10.1016/j.msea.2009.01.002.

16. Z. Zhang, J. Liu, T. Wu, Y. Xie, Effect of carbon nanotubes on friction and wear of a piston ring and cylinder liner system under dry and lubricated conditions, *Friction.* 5 (2017) 147–154. doi:10.1007/s40544-016-0126-6.

17. P. Sharma, S. Sharma, R. Kumar Garg, K. Paliwal, D. Khanduja, V. Dabra, Effect of graphite content on mechanical properties and friction coefficient of reinforced aluminum composites, *Powder Metall. Met. Ceram.* 56 (2017) 264–272. doi:10.1007/s11106-017-9894-4.

18. P. Hariharasakthisudhan, S. Jose, K. Manisekar, Dry sliding wear behaviour of single and dual ceramic reinforcements premixed with Al powder in AA6061 matrix, *J. Mater. Res. Technol.* (2018) 1–9. doi:10.1016/j.jmrt.2018.01.005.

19. S. Sharma, T. Nanda, O.P. Pandey, Effect of particle size on dry sliding wear behaviour of sillimanite reinforced aluminium matrix composites, *Ceram. Int.* 44 (2018) 104–114. doi:10.1016/j.ceramint.2017.09.132.

20. M. Paidpilli, G.K. Gupta, A. Upadhyaya, Effect of matrix powder and reinforcement content on tribological behavior of particulate 6061Al-TiB$_2$ composites, *J. Compos. Mater.* 53 (2019) 1181–1195. doi:10.1177/0021998318796172.

21. S. Poria, P. Sahoo, G. Sutradhar, Tribological characterization of stir-cast aluminium-TiB$_2$ metal matrix composites, *Silicon.* 8 (2016) 591–599. doi:10.1007/s12633-016-9437-5.

22. P.R. Samal, P.R. Vundavilli, A. Meher, M.M. Mahapatra, Influence of TiC on dry sliding wear and mechanical properties of in situ synthesized AA5052 metal matrix composites, *J. Compos. Mater.* (2019) 002199831985712. doi:10.1177/0021998319857124.

23. M. Megahed, M.A. Attia, M. Abdelhameed, A.G. El-Shafei, Tribological characterization of hybrid metal matrix composites processed by powder metallurgy, *Acta Metall. Sin. (English Lett.* 30 (2017) 781–790. doi:10.1007/s40195-017-0568-5.

24. F. Chen, N. Gupta, R.K. Behera, P.K. Rohatgi, Graphene-reinforced aluminum matrix composites: A review of synthesis methods and properties, *Jom.* 70 (2018). doi:10.1007/s11837-018-2810-7.

25. D. Ozyurek, S. Tekeli, A. Gural, A. Meyveci, M. Guru, Effect of Al$_2$O$_3$ amount on microstructure and wear properties of Al–Al$_2$O$_3$ metal matrix composites prepared using mechanical alloying method, *Powder Metall. Met. Ceram.* 49 (2010) 289–294. doi:10.1680/jemmr.16.00016.

5 Energy Materials and Energy Harvesting

K.S. Smaran
Sri Sathya Sai Institute of Higher Learning

S.G. Patnaik
Laboratory for Analysis and Architecture
of Systems: LAAS-CNRS

V. Raman
International Advanced Research Centre for
Powder Metallurgy and New Materials: ARCI

N. Matsumi
Japan Advanced Institute of Science and Technology

CONTENTS

5.1 Introduction 84
5.2 Cathodes 85
5.3 Separators 87
5.4 Anodes 87
5.5 Binders: Properties and Functions 89
    5.5.1 Chemical and Electrochemical Stability 90
    5.5.2 Electrical and Ionic Conductivity 90
    5.5.3 Adherence/Mechanical Stability 91
    5.5.4 SEI Formation and Electrolyte Interaction 92
5.6 Electrolytes 92
    5.6.1 Organic Liquid Non-Aqueous Electrolytes 93
    5.6.2 Ionic Liquids As Liquid/Quasi-Solid Electrolytes 94
    5.6.3 Solid Polymer Electrolytes 94
    5.6.4 Inorganic Solid Electrolytes or Ceramic Electrolytes 96
    5.6.5 Quantification of Ionic Conductivity 97
5.7 Energy Harvesters and Renewable Technologies 98
    5.7.1 Li-Ion Batteries for Photovoltaics 99
    5.7.2 Power Management 100
    5.7.3 Successful Chemistries 101
5.8 Future Market Projections 102
5.9 Conclusions 103
References 103

5.1 INTRODUCTION

Electrochemical energy production or storage has gathered sufficient momentum as sustainable and environmentally benign alternative in satisfying the energy-craving global demands. In this regard, secondary lithium-ion batteries (LiBs) have over shadowed all other contemporary energy devices due to its spectra of applications and ease of usage. The recognition of LiBs over other sources such as lead-acid (LA) or nickel-cadmium (Ni-Cd) or nickel-metal hydride (Ni-MH) was primarily due to its superior energy density, high operating potential range, minimal self-discharge, rapid ability to recharge, and portability [1,2]. The first commercial LiB by Sony boasted of an energy density of 200 Wh/L and 80 Wh/kg [3].

Despite LiBs sharing a great trend in marketability, safety (especially thermal safety due to use of liquid electrolytes) and cost concerns have triggered a need for enhancement of the existing technology. The material scientists including the theoreticians and the experimentalists are on a pursuit for a re-designed version of the LiBs catering core areas of design of materials, cyclic durability, higher safety, and economic production lines [4].

Currently, liquid electrolytes, popularly employed in LiBs, are significantly advantageous due to ionic conductivity, facile interfacial area kinetics with multiple electrode combinations, adaptable with the volume changes of electrodes during charge and discharge cycles. However, liquid electrolytes are extremely flammable and hazardous and have the potential to fuel thermal runaway within batteries [5]. Thermal runaway of liquid-based LiBs can result in an explosion; therefore, it is necessary for such batteries to contain additional safety measures to ensure enclose the liquid safely and to prevent safety issues. Any of such additional inclusions in terms of weight hamper the overall energy density of the batteries. Furthermore, liquid electrolyte batteries typically exhibit solvent leakage and rapid self-discharge, especially at elevated temperatures. Liquid electrolyte-based LiBs are also susceptible to the decomposition of the electrolyte/additives on the electrode materials, which can result in poor performance. Till date, many of the commercially available gadgets powered by LiBs contain these hazardous liquid electrolytes regardless of their safety drawbacks as their performance is unparalleled. In order to construct batteries with a higher degree of safety, liquid electrolytes must be replaced with a viable alternative. Recent research trends hint at solid-state electrolytes as the practical alternative to flammable liquid electrolytes for LiBs [6]. Solid-state electrolytes are advantageous in terms of mechano-electrochemical stability and improved shelf life. Additionally, batteries employing solid electrolytes outclass liquid electrolytes due to non-flammable attributes and thus, nullifying chances of a thermal runaway and therefore, independence from the extra safety inclusions. With only lithium ions are mobile moieties in solid electrolytes, undesirable side reactions and/or decomposition of the electrolytes/ or additives are also highly minimized. Unfortunately, current solid-state electrolytes cannot perform comparably with liquid electrolytes due to lower achievable conductivity and poor interfacial characteristics with the anode. The lithium metal based battery technologies are very promising, owing the attractive features of lithium anode which includes low weight and volume per unit charge, and

Energy Materials and Energy Harvesting 85

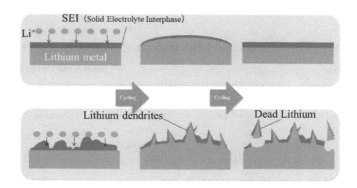

FIGURE 5.1 Illustration depicting SEI formation and dendrites on lithium anode.

high cell voltage. However, the formation of dendritic plating on the lithium anode during the charge discharge cycle leads to safety concerns (Figure 5.1).

The issue of dendritic propagations has been a major deterrent acting against the prospects of the lithium anode. The advent of solid electrolytes has allowed revisiting this concern [7]. Table 5.1 gives a summary of the merits and demerits of liquid electrolytes over all solid-state batteries. The challenge for materials scientists and battery researchers alike is to design and develop a magical battery composition that is safe from flammability concerns, cost-effective in procurement of raw materials, with simultaneous high energy density.

5.2 CATHODES

Cathode is an important component, having a major role in the overall efficient function of the LiBs. This electrode acts as a temporary reservoir of lithium ions, during the cycling of the battery, specifically in the discharged state. In principle, upon charging, the lithium ions move out from the host towards the anode. Upon discharge, the lithium ions are shifted back to the host material, i.e., the cathode. Performance of cathode is dependent on the structural morphology and the microstructures of the cathode material. These parameters critically determine its electrochemical performance. In general, cathode materials have exhibited two pathways of energy storage: (i) Lithium-ion intercalation mechanism and (ii) conversion reaction mechanism. Materials undergoing intercalation pathway, in simple, act as a reversible host for Li ions. Whereas the materials undergoing conversion pathway traverse a solid-state reduction and oxidation reaction during the process. Identification of new materials with good structural stability is of considerable interest.

Among intercalation materials, chalcogenides, transition metal oxides and polyanion compounds have exhibited promising attributes. Transition metal oxides have gathered more interest compared to other intercalating materials owing to its higher voltage window in which the system can be operated, leading to higher energy density. To date, $LiCoO_2$ has been most commonly employed as a cathode in commercial LiBs. The highest achievable voltage due to delithiation from $LiCoO_2$ (~ 4.2 V *vs.* Li/Li$^+$),

TABLE 5.1
Merits and Demerits of the Liquid Electrolytes *vs.* All Solid-State Electrolytes

Liquid Electrolyte	All Solid-State Electrolyte
Inexpensive processing	Expensive processing
Production in large format	Production possible only in small format (to-date)
Ambient tolerance of mechanical stress due to flexible separator	Minimum tolerance of stress due to rigid, ceramic separator
Low interfacial impedance	Higher interfacial impedance
High ionic conductivity at Room Temperature (RT) RT	High ionic conductivity beyond RT
Highly flammable electrolyte, fire prone	Nonflammable electrolyte, very safe
SEI layer forms due to electrolytic decompositions	No SEI layer formation
SEI layer has pronounced effect on cycle life	Cycle life is independent of SEI layer
Electrolyte stability limits the type of cathode material to be used	Electrolyte is nonvolatile, high-voltage cathode materials can used without difficulty
Thermal stability is a concern	Very good thermal stability
Shelf life is limited by self-discharge	Longer shelf life due to less self-discharge
Overcharge protection required	Tolerant to accidental overcharges
Dendrite growth goes unobstructed	Dendrite growth is largely restricted
Reduction in energy density, and specific energy due to high ratio of inactive materials	Increase in energy density, and specific energy due to low ratio of inactive materials

with a resultant maximum practical output of 140 mAh/g. The major advantages of the LiBs have been attributed to these cathode properties, which not only give high capacity but also have relatively long shelf life. $LiCoO_2$ was the first material to be employed as a cathode by the electronic giant, Sony. They exploited this material due to the material's high structural stability and manufacturability. Use of Co in the material has always been the limiting factor in employing $LiCoO_2$ in larger electrodes. As a result, the quest for abundant and efficient material at a lower cost led to layered materials such as $LiNiO_2$. Though $LiCoO_2$ and $LiNiO_2$ share similar layered crystal structures, the inherent higher capacity of $LiNiO_2$ of $\sim 20\%$–30% compared to $LiCoO_2$ gives the advantage. Thus, $LiNiO_2$ by virtue of its capacity is considered as the second-best alternative to $LiCoO_2$.

A third-generation intercalating material by employing ternary transient metal systems of $LiNi_xMn_yCo_zO_2$ (with $x+y+z=1$ or NMC) is the new entrants. In recent years, NMC-based cathodes are replacing $LiCoO_2$ considering the meritorious specific capacity, with an excellent rate capability and cyclability and also for reducing the cobalt quotient in the material composition. Research trends hint at NMC 811 as the most efficient among the NMC family of cathode materials such as NMC 111 or NMC 622. However, NMC 811 undergoes severe structural degradation upon continuous cycling, making it vulnerable to commercial production. In recent years, over-lithiated oxides (OLOs) have shown very high capacity in LiBs and hence a significant research interest has been realized. The OLOs have a general formula $Li_2MnO_{3(1-x)}LiMO_2$. OLOs are most notable as a high voltage cathode material (> 4.5 V $vs.$ Li/Li^+), with practical capacity > 250 mAh/g. However, OLOs face

Energy Materials and Energy Harvesting

a technical issue due to the formation of Li_2O in its initial stages. Spinel oxides are yet another class of cathode materials denoted by the general compositional formula $LiMn_2O_4$ (LMO). Its spinel framework makes it conducive for lithium-ion movement without excessive volume changes. However, LMO faces low compatibility at high-temperature conditions. The maximum practical capacity of LMO reaches up to a minimal 120 mAh/g making it a difficult choice for commercial production. At present composite cathodes comprising NMC and LMO tap the mixed benefits of either of the cathode materials. $LiFePO_4$ (LFP) olivines are polyanionic compounds that are another class of cathodes gaining interest due to optimum thermal stability and power capabilities. In LFP, the structural stability and thus safety are mainly due to the presence of strong P–O bond characteristics, which resists its break down. Due to scarcity of raw materials among the Co/Ni/Mn-based cathodes, LFP is relatively economically viable and thus, an ideal choice for new battery technologies.

A stable crystalline cathode material is an essential prerequisite and should have a wide spectrum of composition in order to suit the charging cycles. Generally, oxidation reactions tend to create huge compositional metamorphism leading to unfavorable phase changes. Hence, materials undergoing conversion pathway experience a change in the crystalline state, which is further accompanied by bond-breaking and bond-forming steps. Examples of conversion materials are metal halides such as FeF_2, CoFe, NiF_2. The major disadvantages of conversion materials are high volume expansion, low electron conductivity, and low reversible capacity issues, which have significantly hindered the utilization of conversion-based cathode materials [8–10].

5.3 SEPARATORS

Separators are an inactive chemical entity, present along with the electrolyte, sandwiched between the electrodes facilitating a channelized passage for lithium ions while avoiding a connection between the electrodes. Their role is highlighted by several attributes such as high chemo-thermo-electrochemical stability covering the range of all other chemical moieties in the battery, a high degree of wettability to solvents or electrolytes, optimum porosity to facilitate lithium diffusivity yet blocking chances of short-circuits. In addition, an optimum thickness provides mechanical strength to avoid interference and diffusion of electrodic components. Nowadays, with an increased focus on thermal safety, the separators are tuned to shut down the lithium-ion channeling in case of elevated temperatures. Typically, a commercial grade separator is of the thickness 20–25 μm. Various compounds of the polymeric nature such as polypropylene, polyethylene, PVDF, PAN, PMMA, polyolefins, etc. are used as separators in LiBs [11,12].

5.4 ANODES

Anodes (negative electrode) in the LiBs are expected to essentially have excellent ratings in the electrochemical aspects such as excellent reversibility during cycling of batteries, optimum volumetric and gravimetric capacities, compatibility against higher potential cathodes, longer cycling life, while being lean on budget [8]. Armand and Touzain first reported the application of graphite as an electrode material [13]. Typically, the graphitic carbons intercalate Li^+ through a systematic multi-step insertion until a final state of LiC$_6$ is reached [14]. Any discussion on anodes is incomplete without discussing solid electrolyte interphase (SEI), a concept brought forward by Peled for the alkali-ion system [15]. Typically, it comprises of organic and inorganic components deposited on the surface of the anode due to the electrolytic decomposition reactions. An ideal SEI layer provides conduction channels for exclusive lithium-ion diffusivity, blocks electron passage, and remains flexible to volume expansion/compression during cycling while retaining an optimum thickness. In short, the SEI layer plays a critical role in providing a passivating cover to the anode, thereby avoiding undue parasitic reactions of electrolyte with the anode [16]. A broad classification of the anodes with representative examples is presented in Table 5.2.

As previously mentioned, graphite as an intercalation host undergoes lithiation and de-lithiation in a sequential fashion with a composition Li$_x$C$_6$ (where, in $x = 0$ or 1, representing either a fully discharged or a fully charged state, respectively) [17]. (Figure 5.2 shows a schematic representation of the SEI formation occurring on the graphite electrode [18]).

TABLE 5.2
Summary of Commonly Used Anode Materials in LiBs

Broad Category	Specific Category	Common Examples	Beneficial Aspects
Intercalation electrodes	Carbonaceous	Graphite, graphene, carbon nanotubes, carbon fibers, porous carbon	Optimum electrochemical range, cost-effective
	Titanium based	TiO$_2$, Li$_4$Ti$_5$O$_{12}$ (LTO)	Longevity, safety, and cost-effective
Conversion electrodes	Mainly transition metal-based oxides/phosphides/sulfides/nitrides	Fe, Ni, Co, Mn, Cu, Cr, Mo, etc.	High capacity
Alloys	Mainly Group-IV metals/metal oxides	Si, Sn, Ge, Pb, and corresponding oxides	High capacity

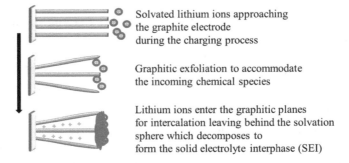

FIGURE 5.2 Schematic representation of the lithium intercalation and SEI formation process in graphite electrode [18].

Energy Materials and Energy Harvesting   **89**

Recently, graphene has gained popularity due to its unmatched mechanical strength and various other superior qualities over graphite. Technically, graphene is a single layer hexagonally arranged sp^2-bonded carbon unlike the planar arrangement of graphite [19]. Despite the high capacity of nanostructured carbon materials as anodes, high surface area of the electrode, accumulation of the nanopores and undesirable functionalization on the electrode surface have resulted in significant irreversible capacity loss. However, surface modification *via* coating processes affects the improvement of retention capacity [20].

Lithium-metal alloys with metals (Si, Ge, Sn) have also been explored as anodes. The theoretical capacity of the Si anode can be as high as 3579 mAh/g at a fully lithiated state of $Li_{15}Si_4$ [21]. However, an anomalous volume expansion along the Si 110 plane has hindered its long-term performance, which had hindered its commercialization [22]. Recently, Silanano® technologies have rolled out commercial Si-anode-based LiBs using nanostructures heavily based on Si [23]. Ge is an expensive Group-IV notable due to superior ionic conductivity and lithium-ion diffusivity. Interestingly, its lithiation leads to an intermediate crystalline state of $Li_{15}Ge_4$ while deintercalation results in an amorphous porous phase with a theoretical capacity up to 1600 mAh/g [24]. Sn-based anodes have a theoretical maximum value of up to 990 mAh/g, which has seldom realized due to undue brittleness of the Sn anode at a fully lithiated state of Li_2Sn_5 [25].

Various metal combinations such as Mg, Al, Ga, and Zn have also been explored as alloying anodes. Magnesium as a prospective anode material has been severely dented due to its passive kinetics, dissolution of Li in Mg, and a strongly passive SEI. Although Al reckons as a good choice due to its economic viability, high theoretical capacity, and low electrochemical potential *vs.* Li/Li$^+$, it has been severely marred due to pulverization of anode during cycling. Ga and Zn have potential attributes such as electrochemically tunable self-healing properties and crystalline phase upon complete lithiation, respectively. However, stable and reversible cycling is yet to be realized in either of the cases [26,27].

5.5 BINDERS: PROPERTIES AND FUNCTIONS

In both anodes and cathodes, active material particles ranging from nano- to micrometer size and conductive additives need to be held together by themselves as well as to the current collector, in order to form the electrode composite laminate. This important function is served by binders, which constitute a small weight fraction of the active mass (> 10%–15% of total weight). Polyvinylidene fluoride (PVDF) has been the industry standard binder for decades until recently when the research focus shifted to alternative high-capacity anodes such as silicon and alloy anodes including tin. The relatively high volume expansion (> 300% in Si and > 280% in Sn) was not compatible with traditional binders such as PVDF, leading to mechanical failure of the electrodes due to loss of contact. Such demands in terms of flexibility, conductivity, chemical and electrochemical stability have thus attracted much research focus in recent years.

TABLE 5.3
Physical and Chemical Constraints for Binder Design

Design Considerations	Desired Properties
Physical characteristics	Soluble in slurry casting solvents
	Insoluble in battery electrolytes
	Extent of swellability in electrolyte
Chemical characteristics	Chemical stability toward Li salts and reduction products
	Electrochemical stability (through HOMO/LUMO calculations
	Favorable interactions with active materials and conductive agents
	No Li-ion blocking effect

5.5.1 CHEMICAL AND ELECTROCHEMICAL STABILITY

Table 5.3 shows various desirable characteristics of binders from practical usage point of view. Since binders form part of the electrode composite, they need to be both chemically and electrochemically stable all through the operating potential window of the battery.

They should be chemically and electrochemically inert to the electrolyte as well as to various electrolyte reduction products such as LiF, Li_2CO_3, etc. at high as well low potentials *vs*. Li. One important and straightforward way to shortlist adequate binders is through computational studies, by calculating their highest occupied molecular orbitals (HOMO) and lowest unoccupied molecular orbitals (LUMO) values. A higher HOMO level indicates the susceptibility of a binder toward oxidation at higher potential, whereas lower LUMO makes it prone to reduction at low potential with respect to lithium.

Figure 5.3 compares the HOMO and LUMO values of different commercially known binders with that of carbonate-based electrolytes [28,29]. Binders such as PVDF, PTFE, and PAN have relatively low HOMO levels and hence can be expected to be electrochemically stable at higher potentials due to the difficulty in removing electrons from outer frontier orbitals. Similarly, the HOMO of PE separator is higher than other polymers and also to that of ethyl carbonate (EC), indicating that it can scarcely remain stable at high potentials. Such initial screening thus allows to predict the electrochemical stability at different potential ranges and also provides essential idea about functional groups that might be changed to obtain desired properties in terms of frontier orbital positions. Also, since the composite electrodes in most cases need to be hot pressed or processed at relatively higher temperatures, the binder behavior at elevated temperatures is also an important factor to consider [30].

5.5.2 ELECTRICAL AND IONIC CONDUCTIVITY

Apart from robust electrochemical stability, the next important factor is that of electrical conductivity and mode of utilization. In the case of conductive binder materials, the need for conductive additives like carbon black is essentially reduced thereby increasing the weight loading of the active material. Several conductive binders have

Energy Materials and Energy Harvesting 91

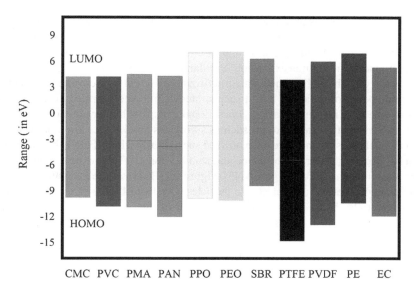

FIGURE 5.3 Comparison of HOMO and LUMO values of different commercially known binders with that of carbonate-based electrolytes [28,29].

been already reported in the literature mostly in anode side with silicon (Si) [31–33] and Si/C [34,35] composites and also with tin (Sn) [8]. However, in most of the cases, it has been seen that the amount of binder required is relatively high (20%–40% of active material). Composite electrodes having conductive additives such as SuperC® and acetylene black for maintaining electrical contact need relatively less amount of binder but have many challenges in maintaining homogenous distribution of all the different components, adherence of active material as well as carbon black particles, etc. [36–39]. In the cathode side, the constraints of high voltage stability and conductivity have led to limited reports [40,41] of successful binder design strategies with scope for further development.

The other important aspect is that of ionic conductivity preventing Li-ion blocking effect and hence ensuring high rate characteristics. This comes from the inherent properties of electrolytes, which have high Li-ion conductivity and their reduction products form polymeric SEI. Inspired from properties of SEI, there have been reports of binders where polymerized electrolyte species such as vinylene carbonate have been utilized as binders showing excellent stability and rate performance even in graphite exfoliating electrolytes like propylene carbonate [42]. Other successful binders like Lipolyacrylate (LiPAA) and also function on similar mechanism having Li-ion conducting channels, which have been shown to function very well with high-voltage cathodes [43].

5.5.3 Adherence/Mechanical Stability

In electrode composites utilizing conductive additives (which is still the case in majority of commercial batteries), binder serves the purpose of adhering the active particles and conductive additives together. Binders having high concentration of carboxyl groups such as carboxymethyl cellulose (CMC), polyacrylic acid (PAA), and alginate-based polymers interact with active materials via hydrogen bonding as well as through covalent bond formation and also have high elastic modulus to efficiently accommodate large volume expansion. Similarly, binder systems having $-NH_2$ groups work based on self-healing mechanism, utilizing different non-covalent interactions and amide bond formation [44]. But such strategies fail when applied in case of cathode side due to oxidation of carboxyl functional groups at higher potentials.

Electrochemically stable groups such as sulfonates and ketones are more appropriate in the cathode side and have been utilized successfully with $LiFePO_4$-based cathode materials showing better performance and stability than PVDF owing to non-covalent interaction from the sulfonate group [45]. Figure 5.4 illustrates the mechanism of binding interactions between the electrode particles with that of the active carbon and the binder materials. There exists feasibility of types of interactions during such binding actions, such as covalent bonds (amide or ester bonds) or non-covalent interactions, such as van der Waals forces or hydrogen bonds or even ion–dipole interactions [46].

5.5.4 SEI Formation and Electrolyte Interaction

Since binders coat the active materials, their functional groups are also the major contributors to the formation of SEI. Also, different binder systems use different processing solvents, leaving behind traces, which also affect the SEI reactions. For example, in lithium titanate (LTO)-based electrodes, PVDF-based electrodes showed rapid accumulation of electrolyte decomposition products, whereas CMC-Na- and PAA-Na-based electrodes showed thinner and more favorable SEI formation. Compared to CMC-Na-, PAA-Na based binder exhibits increased carbonate content due to higher water content from processing [47]. Similar effects of binder systems on SEI formation have also been marked in Si anodes where PAA-CMC composite binder system exhibits better cyclability and SEI formation [48].

5.6 ELECTROLYTES

Electrolytes are sandwiched between the electrodes and constitute the most significant role of transporting the lithium ions between the electrodes and thus, are instrumental in pronouncing the performance of the batteries. The choice of electrolytes

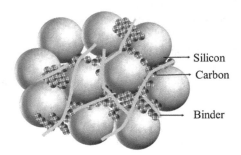

FIGURE 5.4 Illustration depicting the interactive mechanism of binders with Si anodes [46].

Energy Materials and Energy Harvesting 93

becomes critical in the presence of a diverse class of electrode materials. The key features defining an ideal electrolyte are as follows:

- **Retention of the SEI**: Should be able to retain the interfacial structure during electrochemical cycling, which includes a significant amount of volume changes of anode.
- **Ionic conductivity**: Prerequisite parameter of ionic conductivity should be at least in the order mS/cm or higher and a Li-ion transference number closer to unity, which quantifies the fraction of lithium-ion movement has govern a significant contribution.
- **Solvation of the lithium ion**: Ease of lithium salt dissociating from the solvation sheath result in easy intercalation into the anode. Hence, the solvation sphere of the lithium ion decides the diffusive aspects.
- **Wide electrochemical window**: With the discovery of high voltage cathode materials with a potential range $\geq 5V$, electrolytes need to be equally compatible.
- **Operational stability**: High-temperature and low-temperature electronic applications necessitate electrolytes operating within a wider temperature range.
- **Safety aspects**: Mechanical strength and non-flammability or flame-retardant prospects in electrolyte combinations block all possibilities of an inadvertent thermal runaway accidents or short-circuiting [28].

Besides the factors mentioned above, low electronic conductivity, low toxicity levels, and most importantly economic viability are other important factors deciding a battery electrolyte. The discovery of polyethylene oxide (PEO) as electrolyte materials for lithium batteries was a gamechanger, which significantly contributed to the development of the LiBs. A simple model of classification of electrolytes would include organic liquid non-aqueous electrolytes, ionic liquids, solid-state electrolytes (inorganic solid electrolytes (ISEs)/solid polymer electrolytes), and hybrid or composite electrolytes [49–51].

5.6.1 Organic Liquid Non-Aqueous Electrolytes

Essentially, this type of electrolyte solution comprises a solvent and a solute, *viz.*, a lithium salt, respectively. Numerous solvents have been experimented that include acetonitrile, γ-butyrolactone, dimethoxyethane (DME), EC, dimethyl carbonate (DMC), diethyl carbonate (DEC), propylene carbonate (PC), tetrahydrofuran (THF), etc. The solvents have been traditionally explored on the basis of electrochemical and thermodynamic considerations. Similarly, lithium salts are also chosen on the basis of ionic conductivity, solubility, electrochemical parameters, thermal stability, cost efficacy, etc. Some common examples of the lithium salts are $LiPF_6$, $LiBF_4$, $LiN(CF_3SO_2)_3$, $LiClO_4$, $LiAsF_6$, $LiCF_3SO_3$, etc. Choice of lithium salt is governed by factors such as ionic mobility and ease of dissociation, thermal stability, conducive toward an ideal SEI formation, and corrosion susceptibility and most importantly, economic viability [49].

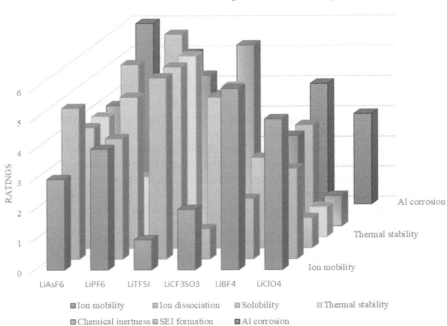

FIGURE 5.5 Chart depicting the properties of various lithium salts [52].

Figure 5.5 charts the properties of various lithium salts [52]. So far, LiPF$_6$ along with a mixture of EC:DMC/DEC has been a viable electrolyte solution commonly used in the commercially available LiBs.

5.6.2 Ionic Liquids As Liquid/Quasi-Solid Electrolytes

Ionic liquids [53] are another category of popular electrolytes for LiBs. Although commercial aspects have been severely limited due to economic considerations, they provide a desirable combination of the presence of a discrete combination of designable cations and anions. The common cations employed as ionic liquids include functionalized derivatives of an aliphatic family such as piperidinium, pyrrolidinium, quaternary ammonium or phosphonium, aromatic type (imidazolium, pyridinium), or pyrrolidinium in the presence of numerous anions such as FSI$^-$, TFSI$^-$, PF$_6^-$, BF$_4^-$ to name a few [54]. Similarly, several categories of polymeric ionic liquids have been employed as electrolytes [55].

5.6.3 Solid Polymer Electrolytes

This category encompasses a wide variety of materials, which include any kind of polymer in the electrolyte matrix. In fact, a few of commercial LiBs with polymer electrolyte are also known as Li-Po batteries. A generalized classification would be

Energy Materials and Energy Harvesting

in the form of dry solvent-free polymer electrolyte, plasticized gel polymer electrolytes, and composite polymer electrolytes. Solvent-free polymer electrolytes comprise polymers of ethers, imines, esters, or even copolymers based on polysiloxanes, polyphosphazenes, etc. These polymeric matrices act as dry solvent for a suitably dissociable lithium salt as the solute. For instance, an easily dissociable lithium salt with a large anion stabilized by delocalized negative charge would be preferable choice over conventional halides as counter anions. Some examples of such delocalized anions are ClO_4^-, $CF_3SO_3^-$, $(CF_3SO_2)_2N^-$, BF_4^-, BPh_4^-, AsF_6^-, SCN^-, etc. Armand et al. reported the first use of a polyethylene oxide chains PEO-LiX, where in PEO chains were utilized as polymer host matrices [56–59]. A set of commonly used polymers are shown in Table 5.4 [60]. These polymeric matrices by virtue of the segmental motions of the intra- and inter-chains result in the ionic conductivity, mostly in the amorphous state. This requirement had given rise to the introduction of plasticizers or solvents or inorganic fillers to enhance the conductivity. Plasticizers, enhance ionic conductivity by reduction in the crystallinity of the parent matrix besides enriching the segmental motions of the polymers. Some commonly employed plasticizers are polyethylene glycol (PEG) and its alkoxy forms, polypropylene glycol (PPG), crown ethers, alkyl carbonates like EC, PC, DMC, DMF *etc.* A clear cut demarcation between gel electrolytes and plasticized electrolytes remains unclear. Hence, many researchers have overlapping views on this matter. Gel polymer electrolytes represent quasi-liquid and quasi-solid electrolyte-like behaviors. The use of separator or flammability concerns can be negated in these electrolytes. A variety of gel polymer electrolytes have been explored with ionic liquids as plasticizers. In this regard, ionogels as electrolyte materials constitute compositions with ionic liquids that are dispersed in an inorganic matrix to mimic the requirements of a polymer electrolyte while adding extra conductivity features due to ionic liquids [61]. Similarly, composite polymer electrolytes (CPEs) are another broad category of electrolytes, which are populated by the incorporation of various chemical moieties like fillers such as TiO_2, SiO_2, Al_2O_3, and $ZnAl_2O_4$ [62]. Anion-trapping by incorporation of boron moiety has been a scientifically proven concept using boroxine ring receptors [63] or alkyl/aryl borates [64] or esters [65] or other organoborane reagents, which regulates the flow of anion at a molecular level [66].

TABLE 5.4
Commonly Used Polymer Matrices Employed in Electrolytes

Polymer Host	Repeat Unit
Poly(ethylene oxide)	$-(CH_2CH_2O)_n-$
Poly(propylene oxide)	$-(CH(-CH_3)CH_2O)_n-$
Poly(acrylonitrile)	$-(CH_2CH(-CN))_n-$
Poly(methyl methacrylate)	$-(CH_2C(-CH_3)(-COOCH_3))_n-$
Poly(vinyl chloride)	$-(CH_2CHCl))_n-$
Poly(vinylidene fluoride)	$-(CH_2CF_2)_n-$
Poly(vinylidene fluoride-hexafluoroproylene)	$-(CH_2CH)-$

5.6.4 Inorganic Solid Electrolytes or Ceramic Electrolytes

ISEs or ceramic electrolytes (CEs) are notably high elastic moduli and have greater thermal stability and brittle. They are composed of mobile ions arranged into a polyhedral coordination with ligands in a solid-state matrix in the crystal form. Crystalline electrolytes register enhanced thermal stability over the amorphous/glass counterparts. Unlike liquid electrolytes, where solvated ion sheath moving through the electrolyte solution forms the ion conduction pathway, ISEs depend on a pathway defined by energetic barriers between crystallographic states. Thus, in most of the ISEs/CEs, vacancies and interstitial grain boundaries are the limiting factors for ionic conductivity. Taking advantage of the proportional relationship of ionic conductivity and temperature, such electrolytes are ideal for high-temperature applications [67,68]. Figure 5.6 shows a comparative plot of the ISE *vs.* commercial electrolyte [69].

There are various subcategories within this group, such NASICON, LISICON, Perovskites, Garnets, etc. To start with, NASICON abbreviated from sodium Na Super Ionic CONductor, (Common formula: $NaM_2(PO_4)_3$, M=metal ion) refers to a class of sodium super ion conductors. A lithium variant of this class results in $LiM_2(PO_4)_3$, a lithium super ion conductor. These conductors typically exhibit monoclinic structure. Incorporation of metals such as Ti, Ge, Hf, Zr has resulted in significant improvements in terms of conductivity and compatibility with electrodes. Also doping of metals such as Al and Fe to create a mixed ratio of metal additions has significantly contributed to the enhancement of its ionic conductivity. An addition of lithium salt such as Li_3BO_3 or $Li_3(PO_4)_3$ has resulted in the densification of the electrolyte, which is known to influence ionic conductivity of such electrolytes [70]. LISICON is a category of lithium super ionic conductor, with a chemical formula $Li_{14}Zn(GeO_4)_4$ when it was first reported. The structural matrix of the electrolyte bears similarity with γ-$Li_3(PO_4)_3$. However, despite the structural similarity of γ-$Li_3(PO_4)_3$ with LISICON, the mechanism of ionic conduction varies largely due to interstitial vacancies for Li-ion hopping, still the order of conductivity remains low for commercialization [71]. Typical examples of this category comprise Li_2S-P_2S_5 glass or glass ceramic, determined by the phase composition. Typically, in this category of electrolytes, the conductivity is proportional to the ratio of crystalline phases. Thio-LISICONs, a variant denoted

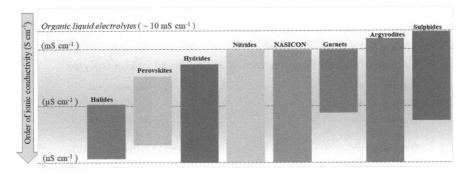

FIGURE 5.6 Depiction of the conductivity parameters of inorganic solid-state electrolytes against a commercial liquid electrolyte (1 M $LiPF_6$ EC:DMC) [69].

Energy Materials and Energy Harvesting 97

by the general formula $Li_2SGeS_2P_2S_5$, show conductivity closer to liquid electrolytes with a negligible electronic conductivity, beside significant thermal stability up to 500°C [72]. Sulfides as electrolytes are another excellent category represented by the general formula, $Li_{10}GeP_2S_{12}$, which are notably high electronic conductivity aided by high mechanical strength. However, hygroscopic nature of such sulfides affects their atmospheric stability and electrochemical stability [73]. Perovskites, with general formula (LaLi)TiO$_3$ (LLTO), are one of the category of oxide solid-state electrolyte. The control of the ratio of La/Li in the crystal structure is critical in determining the ionic conductivity parameters. Incorporation of oxides such as Li_2O, Li_3BO_3, Li_4SiO_4, etc. or inert oxides such as SiO_2 or Al_2O_3 has resulted in the improvement of grain boundary conductivity [74]. Lithium-rich anti-perovskites are technically an "electronically inverted" perovskites with a general formula of LiO_3Cl. Similar to the perovskites, this class of superionic conductors possesses excellent conductive parameters, beside a low activation energy in the range of 0.2–0.3 eV [75,76]. Garnet-type oxides are represented by a general formula of $Li_5La_3M_2O_{12}$ (M=Nb, Ta). LLZO type $Li_7La_3Zr_2O_{12}$ as a promising solid electrolyte for lithium batteries. Similarly, ternary-doped LLZO has enhanced ionic conductivity due to dense microstructure and resultant increased lithium-ion occupancy in the crystal structure [77–80]. Table 5.5 presents a comprehensive list of the material properties of the ISEs [81].

5.6.5 Quantification of Ionic Conductivity

In any category of electrolytes, the patterns of ionic conductivity are quantified and explained by the two models: (i) Arrhenius model – ideal in cases with identified crystal lattices, given by the Equation 5.1:

$$\sigma = \sigma_o e^{\left(\frac{-E_a}{kT}\right)} \text{ or } \sigma = A \tag{5.1}$$

where σ_0 or A=the carrier ion number, E_a=activation energy, k=Boltzmann constant, and T=absolute temperature in K. A linear plot of ln σ vs $10^3/T$ indicates a typical Arrhenius behavior with no occurrence of phase transition.

TABLE 5.5

A Comprehensive List of the Material Properties of the ISEs [75]

Examples of ISE	Chemical Stability	Electrochemical Stability	Mechanical Strength	Grain Boundary Resistance	Conductivity	Manufacturing Feasibility
NASICON						
LISICON	+++	+++	+++	+	+++	+
Perovskites						
Garnet						
Sulfides	+	+	++	+	+++	+
Anti-perovskites	+	+	++	++	+	+

Functional and Smart Materials

Another advanced model depicting ionic transport is the Vogel Tammann Fulcher (VTF) model, based on free volume theory, accounts for more accurate representation of the segmental motion of the polymer electrolytes, through Equation 5.2:

$$\sigma = \frac{A}{\sqrt{T}} e \tag{5.2}$$

where A, σ and T are same as Equation 5.1, B refers to the pseudo-activation energy, while T_0 is equilibrium glass-transition temperature, where T_0 is roughly 50 K lower than T_g [82].

5.7 ENERGY HARVESTERS AND RENEWABLE TECHNOLOGIES

An energy harvester can be defined as a system that produces energy from its surrounding without consuming any fuel. Ambient energy sources can be classified into three broad categories: thermal, kinetic, and electromagnetic sources. Table 5.6 summarizes various sources and relevant technologies for harvesting them.

Renewable technologies have been developing at rapid pace owing to various public incentives and policies and will soon become mature and competitive to completely remove the dependence on fossil fuels. Such developments have made independent power providers to come up with various non-conventional technologies, which are emission free and efficient, adding up to the overall grid reliability and independence. The dependable structural basis sets for graduating to renewable resources based intelligent and smart grids are mostly the "micro grids." Such micro grids not only offer better flexibility but also security, efficiency in all stages of overall grid functioning. Micro grids are miniature grids, which can function, connected to the main grid or autonomously, comprising of various distributed energy resources (small-scale energy generation and storage such as photovoltaics, batteries, fuel cells, micro turbines, etc.) [83]. Even though such systems might appear to be promising, their ability

TABLE 5.6

Classification of Various Sustainable Energy Sources and Current Harvesting Technologies

Type of Energy	Source	Harvesting Technologies
Thermal	Natural: Any kind of available temperature gradient	Thermo electric generators using thermoelectric materials
	Industrial: Furnaces, heaters, chillers, fans, friction sources, car engines, etc.	
Kinetic/mechanical	Natural: Wind, water flow, ocean currents, body movements, etc.	Piezoelectric generators
		Electrostatic generators
	Industrial: Mechanical vibrations, stresses and strains from machines	Electromagnetic generators
Electromagnetic radiation	Natural: EM waves, RF signals, sunlight, etc.	Solar panels/ Photovoltaic transducers, RF transducers
	Industrial: Inductors, coils, transformers, sun light, etc.	

Energy Materials and Energy Harvesting 99

is severely limited by inconsistency of various renewable resources such as wind and solar, and by utilization of inverter interfaced power generators. This kind of situation thus opens up a niche for a variety of energy storage systems which can be tuned up for serving the evolving grid architecture [84,85].

5.7.1 Li-Ion Batteries for Photovoltaics

Batteries can enable storage at desired time and locale – wedging apart energy generation and its expenditure, thus bringing more stability and security. Among various available battery technologies, Li-electrochemistry has been distinctly advantageous in these applications owing to its extremely low weight (high gravimetric capacity), low redox voltage w.r.t normal hydrogen electrode (NHE) (enabling high output voltage and electrochemical stability) and small ionic radius (easy diffusion) [86]. Nevertheless, for graduating to wide-scale grid-level application, there is still a lot of scope for development in terms of life cycle, energy density, safety, cost reduction, and rate kinetics. Moreover, contrary to automotive and consumer electronics, the battery technology evolution for sustainable grids needs to follow a modular approach, where focus can be on standardization for allowing flexible arrangement and easy construction for each specific sub-application in the grid [87–89]. Among different renewable energy sources, solar energy is the most widespread. The solar energy received by earth covers a wide spectrum and is ~ 1370 W/m², making it the most abundant renewable energy source. Hence, in the current section, discussion is limited to the application of Li electrochemistry to photovoltaic systems as an example to draw the attention of the reader toward the role of LiBs in energy harvesting.

Figure 5.7 displays the contribution batteries in conjugation with different components of the grid system. Unlike in portable electronics, batteries in renewable energy harvesting form a part of broader installation, thereby requiring additional control systems to address the intermittency of solar power and may be utilized for different specific roles. In small-scale PV installations, drastic power fluctuations

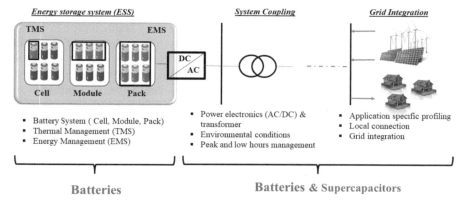

FIGURE 5.7 Schematic showing contribution of batteries and supercapacitors in different levels of grid functioning.

(due to clouds or weather change) not only affect the reliable power supply but also can reduce deliverable output ramp rate required by the system operator. Even after ramp rate control, intermittency of solar power can also cause occasional frequency deviation anywhere up to 2–3 Hz and provided the fact that different PV facilities are interconnected, frequency change in the order of ~0.1 Hz can cause serious issues. In case of large-scale installations having diversity of generation sources which are interconnected, ramp rate control and frequency change may not be an issue, but rapid change in power will be a cause of concern for providing steady output voltage to end users. Battery energy storages (BESs) can thus answer issues faced by varies kinds of installations by acting as an interface between the generation source and the end user.

BES can be integrated with PV systems at two main levels. Firstly, a centralized storage system (in MW scale) at the distribution station near to the PV facility to reduce complications in control systems and ease of access to the PV facility, thereby leading to efficient and high-quality power dispatch from the source itself. This kind of centralized storage thus needs large space, huge installation set up, a dedicated on-site technological support. This kind of integration is thus well suited for *large-scale installations*. The other option can be to install small-scale BESs at different points in the distribution line, depending on the need and application, but all networked together and remotely operated at a centralized PV facility. This provides the advantage of need based energy dispatch at the point of consumption. However, such a system requires complicated networking, technology, and work force to monitor and maintain individual consumption sites and hence well suited for *small-scale installations*. However, depending upon the need and the specific application, one needs to choose the most appropriate battery chemistry to ensure safety, reliability, and low maintenance. This stems from the inherent limitation of different available battery chemistries. For example, BES for frequency regulation application in the grid, needs high-power capability to address grid fluctuations but experiences moderate cycle duty (rate below 0.5 C) depth of discharge (DOD) seldom exceeds ±20%. This can lead to accelerated aging of typical Li ion cells. But in case of BES for peak saving type applications in PV systems, it needs to survive high C rate (1.2% of the time with C rate ≥ 1, peak at 3.5 C) [90]. These variable requirements thus need tailored solution strategies for individual modules of energy storage in a grid, opening up the scope for significant development in different battery technologies.

Even though mature, LiBs are still quite expensive for wide-scale application in grid-connected PV systems, where lead-acid battery chemistry (even though not as efficient as LiB) dominates due to cost factor. However, in small-scale and residential PV applications, LiBs are the best suiting technology if appropriate electrode chemistries are chosen. Apart from having high depth of discharge, long cycling stability (minimum >5000 cycles), cost-effective production and transport, moderate to slightly high-temperature operation are also governing factors for choosing the appropriate materials.

5.7.2 Power Management

The power management circuitry bridges the energy harvesting and energy storage units. The energy input from the harvesters is in the form of instantaneous pulse with hundreds of volts, μA level current, and MΩ or higher resistance. The energy

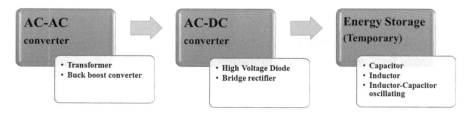

FIGURE 5.8 Three main components of the power management circuit.

required for charging the storage devices in the range of few volts (<3 V for supercapacitors and <5 V for batteries) and <kΩ impedance. Hence, in this scenario, the role of the power management circuit is to efficiently manage the electric energy and transfer to the energy storage devices. It comprises of three main components as shown in Figure 5.8.

The AC-AC converters boost the output current at the cost of the output voltage. Similarly the AC-DC converters covert the AC to DC signal for the energy storage devices. This is followed by temporary energy storage in capacitors or inductors which is further connected to long-term energy storage systems or to user terminals. However, as different battery chemistries allow different rate and extent of charging, it is essential for the BMSs (battery management systems) to prevent overcharging or over-discharging which might affect the health of such devices. In modern day BMS, it is usually done by monitoring the state of charge (SOC) of the batteries. Ease of monitoring the SOC in a given battery chemistry is thus one of the important criteria for the utilization of a given device setup.

SOC is defined as the amount of charge left in the battery compared to its fully charged state. In renewable energy applications such as PV systems, it is very essential to keep the batteries under partial state of charge (PSOC). This necessity stems from the inherent nature of renewable sources so that the battery can either absorb charge or discharge into the grid system at any time, depending on the energy source. For maintaining such an optimal situation, BMS monitors the SOC of the battery.

In commercial applications, the SOC of a battery is measured by monitoring the open circuit potential of the battery. In many battery chemistries, the cell potential drops when the SOC decreases which can be monitored and managed by simple algorithms in the BMS. However, when the cell chemistries have a flat voltage profile such as LiFePO$_4$ (LFP) cathode over LiCoO$_2$ (LCO) or Li$_2$TiO$_3$ (LTO), voltage-based SOC determination becomes difficult, thereby making it difficult for integration with power electronic systems. It is also imperative here to note that drastically large variations of the potential will also be an issue, needing complex circuit systems for system integration [91–94].

5.7.3 Successful Chemistries

In the past decade, there have been growing demands for batteries for PV systems, and many different battery manufacturers have come up with innovative technologies to cater the needs of this growing sector. Table 5.7 shows different commercialized

TABLE 5.7
Details of Commercialized Battery Chemistries Suitable for Energy Storage Systems in Renewable Energy Application

Parameter	Unit	Li B Cell Characteristic				
Cell name		SD194Ah	NCR18650B	US26650FTC1	SCiB Titanate	ANR26650M1-B
Company		Samsung	Panasonic	Murata	Toshiba	A123 Systems
Active materials		NMC/C	NCA/C	LFP/C	MO_x/LTO	LFP/C
Format		Prismatic	Cylindrical	Cylindrical	Prismatic	Cylindrical
Capacity	Ah	94	3.2	3.0	20	2.5
Energy density	Wh/L	355	676	278	177	200
Rate	C-Rate	3C/1C	2C/0.5C	6C/1C	8C/3C	1C/0.5C
Cyclability		>5000	300	>6000	>10,000	>1000
Voltage Range	V	2.7–4.1	2.5–4.2	2.0–3.6	1.5–2.7	2.5–3.6
Average Potential	V	3.7	3.6	3.2	2.3	3.6

chemistries [81,95,96]. Tesvolt recently came up with their award winning BES utilizing $LiFeMnPO_4$ cathodes and carbon-based anodes in prismatic cell set up (which are more durable and safe compared to round cells produced by Samsung electronics) specifically designed for peak-saving applications. Their packs have excellent lifetime of 30 years (8000 cycles) with ability to provide continuous power supply at 1C and up to 4C for short durations. As a point of fact, it is also one of the successful chemistries without utilizing cobalt in their active material. With increasing penetration of renewable sources and greater demand for various BES, it is imperative to ensure sustainability. As a result, research focus should be shifted to more sustainable alternatives for active materials while simultaneously working on technology to make the manufacturing process economic and cost effective.

5.8 FUTURE MARKET PROJECTIONS

With an increased dependence on lithium battery for myriads of applications coupled with increasing environmental concerns, the lithium-ion and its superior variants are going to be in the limelight. The increasing outreach of electric vehicles (EVs) is another area where LiBs are in the perfect niche. The materials development period of around 15 years has been projected for the full-fledged utilization of all-solid-state battery that has been depicted in Figure 5.9.

The battery-type market shares in EV transportation sector have been projected to arrive nearly 90%–100% by all-solid-state batteries by 2030 in a separate market research group [97]. However, corroborating the projections of various data analysis and market statistics groups, it is interesting to grasp that the commencement of all-solid-state-battery era is forecasted from 2030 in all forms of the battery market.

Energy Materials and Energy Harvesting

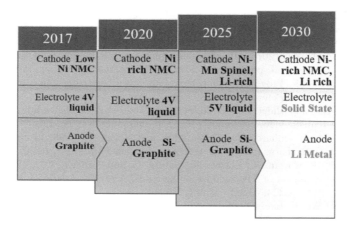

FIGURE 5.9 Materials development projections toward all-solid-state-battery market by 2030 [97].

5.9 CONCLUSIONS

Technological evolution has been on a roll with incessant energy demands, with specifications for more power, less space, and "safe batteries." LiBs have been leading the battery technology for over three decades, while co-evolving and also complementing other juvenile battery technologies. The current pre-requisites of a sustainable technology include not only a viable by-product of scientific innovation but also which economically caters to the market demands of the world. The growing popularity of EVs makes it imminent for the "survival of the fittest" technology. With Tesla making enormous investments in developing Gigafactories, it is evident that the life of LiBs is not ephemeral. Further, environmental concerns make it inevitable for profitable energy harvesting techniques, which so far had been left completely ignored. With scientific research focused on material development for improved output and increased portability, the LiBs are just getting better. Although the prospects of solid-state batteries are promising, the gap between laboratory scale fundamental research translating into commercial products is still very huge. In this regard, the design of task-specific energy materials for the dual goals of energy storage and harvesting, utilizing simulation chemistry, and then realizing it through experimental research might pave the way for a research-intensive, transdisciplinary, and cost-effective world of the future.

REFERENCES

1. Tarascon, J.M. and Armand, M., 2001. Issues and challenges facing rechargeable lithium batteries. *Nature, 414*, p. 359.
2. Armand, M. and Tarascon, J.M., 2008. Building better batteries. *Nature, 451*(7179), p. 652.
3. Nishi, Y., 2001. Lithium ion secondary batteries; past 10 years and the future. *Journal of Power Sources, 100*(1–2), pp. 101–106.

4. Li, M., Lu, J., Chen, Z. and Amine, K., 2018. 30 years of lithium-ion batteries. *Advanced Materials*, *30*(33), p. 1800561.
5. Schnell, J., Günther, T., Knoche, T., Vieider, C., Köhler, L., Just, A., Keller, M., Passerini, S. and Reinhart, G., 2018. All-solid-state lithium-ion and lithium metal batteries–paving the way to large-scale production. *Journal of Power Sources*, *382*, pp. 160–175.
6. Takada, K., 2018. Progress in solid electrolytes toward realizing solid-state lithium batteries. *Journal of Power Sources*, *394*, pp. 74–85.
7. Lotsch, B.V. and Maier, J., 2017. Relevance of solid electrolytes for lithium-based batteries: A realistic view. *Journal of Electroceramics*, *38*(2–4), pp. 128–141.
8. Winter, M., Besenhard, J.O., Spahr, M.E. and Novak, P., 1998. Insertion electrode materials for rechargeable lithium batteries. *Advanced Materials*, *10*(10), pp. 725–763.
9. Etacheri, V., Marom, R., Elazari, R., Salitra, G. and Aurbach, D., 2011. Challenges in the development of advanced Li-ion batteries: A review. *Energy & Environmental Science*, *4*(9), pp. 3243–3262.
10. Nitta, N., Wu, F., Lee, J.T. and Yushin, G., 2015. Li-ion battery materials: Present and future. *Materials Today*, *18*(5), pp. 252–264.
11. Xiang, Y., Li, J., Lei, J., Liu, D., Xie, Z., Qu, D., Li, K., Deng, T. and Tang, H., 2016. Advanced separators for lithium-ion and lithium–sulfur batteries: A review of recent progress. *ChemSusChem*, *9*(21), pp. 3023–3039.
12. Lee, H., Yanilmaz, M., Toprakci, O., Fu, K. and Zhang, X., 2014. A review of recent developments in membrane separators for rechargeable lithium-ion batteries. *Energy & Environmental Science*, *7*(12), pp. 3857–3886.
13. Armand, M. and Touzain, P., 1977. Graphite intercalation compounds as cathode materials. *Materials Science and Engineering*, *31*, pp. 319–329.
14. Aurbach, D., Markovsky, B., Weissman, I., Levi, E. and Ein-Eli, Y., 1999. On the correlation between surface chemistry and performance of graphite negative electrodes for Li ion batteries. *Electrochimica acta*, *45*(1–2), pp. 67–86.
15. Peled, E., 1979. The electrochemical behavior of alkali and alkaline earth metals in nonaqueous battery systems—the solid electrolyte interphase model. *Journal of The Electrochemical Society*, *126*(12), pp. 2047–2051.
16. Verma, P., Maire, P. and Novák, P., 2010. A review of the features and analyses of the solid electrolyte interphase in Li-ion batteries. *Electrochimica Acta*, *55*(22), pp. 6332–6341.
17. Cheng, F., Liang, J., Tao, Z. and Chen, J., 2011. Functional materials for rechargeable batteries. *Advanced Materials*, *23*(15), pp. 1695–1715.
18. An, S.J., Li, J., Daniel, C., Mohanty, D., Nagpure, S. and Wood III, D.L., 2016. The state of understanding of the lithium-ion-battery graphite solid electrolyte interphase (SEI) and its relationship to formation cycling. *Carbon*, *105*, pp. 52–76.
19. Raccichini, R., Varzi, A., Passerini, S. and Scrosati, B., 2015. The role of graphene for electrochemical energy storage. *Nature Materials*, *14*(3), p. 271.
20. Lu, J., Chen, Z., Pan, F., Cui, Y. and Amine, K., 2018. High-performance anode materials for rechargeable lithium-ion batteries. *Electrochemical Energy Reviews*, *1*(1), pp. 35–53.
21. Cho, J., 2010. Porous Si anode materials for lithium rechargeable batteries. *Journal of Materials Chemistry*, *20*(20), pp. 4009–4014.
22. Lee, S.W., McDowell, M.T., Choi, J.W. and Cui, Y., 2011. Anomalous shape changes of silicon nanopillars by electrochemical lithiation. *Nano Letters*, *11*(7), pp. 3034–3039.
23. https://silanano.com/.
24. Kennedy, T., Mullane, E., Geaney, H., Osiak, M., O'Dwyer, C. and Ryan, K.M., 2014. High-performance germanium nanowire-based lithium-ion battery anodes extending over 1000 cycles through in situ formation of a continuous porous network. *Nano Letters*, *14*(2), pp. 716–723.

Energy Materials and Energy Harvesting

25. Winter, M. and Besenhard, J.O., 1999. Electrochemical lithiation of tin and tin-based intermetallics and composites. *Electrochimica Acta, 45*(1–2), pp. 31–50.

26. Obrovac, M.N., Christensen, L., Le, D.B. and Dahn, J.R., 2007. Alloy design for lithium-ion battery anodes. *Journal of The Electrochemical Society, 154*(9), pp. A849–A855.

27. Nitta, N. and Yushin, G., 2014. High-capacity anode materials for lithium-ion batteries: Choice of elements and structures for active particles. *Particle & Particle Systems Characterization, 31*(3), pp. 317–336.

28. Maleki, H., Deng, G., Kerzhner-Haller, I., Anani, A. and Howard, J.N., 2000. Thermal stability studies of binder materials in anodes for Lithium-ion batteries. *Journal of the Electrochemical Society, 147*, pp. 4470–4475..

29. Choi, N.S., Ha, S.Y., Lee, Y., Jang, J.Y., Jeong, M.H., Shin, W.C. and Ue, M., 2015. Recent progress on polymeric binders for silicon anodes in lithium-ion batteries. *Journal of Electrochemical Science and Technology, 6*(2), pp. 35–49.

30. Zhao, H. et al. 2015. Conductive polymer binder for high-tap-density nanosilicon material for lithium-ion battery negative electrode application. *Nano Letters, 15*, pp. 7927–7932.

31. Wu, H. et al. 2013. Stable Li-ion battery anodes by in-situ polymerization of conducting hydrogel to conformally coat silicon nanoparticles. *Nature Communications, 4*, p. 1943.

32. Wu, M. et al. 2013. Toward an ideal polymer binder design for high-capacity battery anodes. *Journal of the American Chemical Society, 135*, pp. 12048–12056.

33. Zhao, H., Du, A., Ling, M., Battaglia, V. and Liu, G., 2016. Conductive polymer binder for nano-silicon/graphite composite electrode in lithium-ion batteries towards a practical application. *Electrochim. Acta, 209*, pp. 159–162.

34. Park, S. J. et al. 2015. Side-chain conducting and phase-separated polymeric binders for high-performance silicon anodes in lithium-ion batteries. *Journal of the American Chemical Society, 137*, pp. 2565–2571.

35. Xun, S., Song, X., Battaglia, V. and Liu, G. 2013. Conductive polymer binder-enabled cycling of pure Tin nanoparticle composite anode electrodes for a Lithium-ion battery. *Journal of the Electrochemical Society, 160*, pp. A849–A855.

36. Zheng, H., Yang, R., Liu, G., Song, X. and Battaglia, V.S. 2012. Cooperation between active material, polymeric binder and conductive carbon additive in lithium ion battery cathode. *The Journal of Physical Chemistry C, 116*, pp. 4875–4882.

37. Chou, S., Pan, Y., Wang, J., Liu, H. and Dou, S. 2014. Small things make a big difference: Binder effects on the performance of Li and Na batteries. *Physical Chemistry Chemical Physics, 16*, pp. 20347–20359.

38. Amin-Sanayei, R. and Heinze, R. November 2009. *Advantages of High Molecular Weight PVDF Binder in Lithium Ion Cells* - Presented in NASA Aerospace Battery Workshop, Alabama.

39. Yuca, N., Zhao, H., Song, X., Dogdu, M. F., Yuan, W., Fu, Y., Battaglia, V.S., Xiao, X. and Liu, G. 2014. A systematic investigation of polymer binder flexibility on the electrode performance of lithium-ion batteries. *ACS Applied Materials & Interfaces, 6*, pp. 17111–17118.

40. Barsykov, V. and Khomenko, V. 2010. The influence of polymer binders on the performance of cathodes for Lithium-ion batteries. *The Scientific Journals of Riga Technical University, 21*, pp. 67–71.

41. Choi, N.S., Lee, Y.G. and Park, J.K. 2002. Effect of cathode binder on electrochemical properties of lithium rechargeable polymer batteries. *Journal of Power Sources, 112*, pp. 61–66.

42. Zhao, H., Zhou, X., Park, S.-J., Shi, F., Fu, Y., Ling, M., Yuca, N., Battaglia, V. and Liu, G. 2014. A polymerized vinylene carbonate anode binder enhances performance of lithium-ion batteries. *Journal of Power Sources, 263*, pp. 288–295.

43. Pieczonka, N.P.W., Valentina, B., Ziv, B., Leifer, N., Dargel, V., Aurbach, D., Kim, J.-H., Liu, Z., Huang, X., Krachkovskiy, S.A., Goward, G.R., Halalay, I., Powell, B.R. and Manthiram, A. 2015. Lithium polyacrylate (LiPAA) as an advanced binder and a passivating agent for high-voltage Li-ion batteries. *Advanced Energy Materials, 5*(23), p.1501008.

44. Wang, C., Wu, H., Chen, Z., McDowell, M.T., Cui, Y. and Bao, Z. 2013. Self-healing chemistry enables the stable operation of silicon microparticle anodes for high-energy lithium-ion batteries. *Nature Chemistry, 5*, pp. 1042–1048.

45. Chiu, K.-F., Leu, H.-J., Su, S.-H. and Wu, C.-H. 2016. Lithiated and sulfonated poly (ether ether ketone) binders with high rate capability for $LiFePO_4$ cathodes. *ECS Transactions, 73*, pp. 19–26.

46. Choi, J.W. and Aurbach, D. 2016. Promise and reality of post-lithium-ion batteries with high energy densities. *Nature Reviews Materials, 1*, p. 16013.

47. Nordh, T., Jeschull, F., Younesi, R., Koçak, T., Tengstedt, C., Edström, K. and Brandell, D. 2017. Different shades of $Li_4Ti_5O_{12}$ composites: The impact of the binder on interface layer formation. *ChemElectroChem, 4*, pp. 2683–2692.

48. Nguyen, C.C., Yoon, T., Seo, D.M., Guduru, P. and Lucht, B.L., 2016. Systematic investigation of binders for silicon anodes: Interactions of binder with silicon particles and electrolytes and effects of binders on solid electrolyte interphase formation. *ACS Applied Materials & Interfaces, 8*, pp. 12211–12220.

49. Goodenough, J.B. and Kim, Y., 2009. Challenges for rechargeable Li batteries. *Chemistry of Materials, 22*(3), pp. 587–603.

50. Goodenough, J.B. and Park, K.S., 2013. The Li-ion rechargeable battery: A perspective. *Journal of the American Chemical Society, 135*(4), pp. 1167–1176.

51. Xu, K., 2014. Electrolytes and interphases in Li-ion batteries and beyond. *Chemical Reviews, 114*(23), pp. 11503–11618.

52. Chagnes, A. and Swiatowska, J., 2012. *Electrolyte and Solid-Electrolyte Interphase Layer in Lithium-ion Batteries.* In *Lithium ion Batteries - New Developments* Belharouak I. (Ed.) InTech. London. pp. 145–172.

53. Xu, W., Cooper, E.I. and Angell, C.A., 2003. Ionic liquids: Ion mobilities, glass temperatures, and fragilities. *The Journal of Physical Chemistry B, 107*(25), pp. 6170–6178.

54. MacFarlane, D.R., Tachikawa, N., Forsyth, M., Pringle, J.M., Howlett, P.C., Elliott, G.D., Davis, J.H., Watanabe, M., Simon, P. and Angell, C.A., 2014. Energy applications of ionic liquids. *Energy & Environmental Science, 7*(1), pp. 232–250.

55. Osada, I., de Vries, H., Scrosati, B. and Passerini, S., 2016. Ionic-liquid-based polymer electrolytes for battery applications. *Angewandte Chemie International Edition, 55*(2), pp. 500–513.

56. Xue, Z., He, D. and Xie, X., 2015. Poly (ethylene oxide)-based electrolytes for lithium-ion batteries. *Journal of Materials Chemistry A, 3*(38), pp. 19218–19253.

57. Mindemark, J., Lacey, M.J., Bowden, T. and Brandell, D., 2018. Beyond PEO—alternative host materials for Li^+-conducting solid polymer electrolytes. *Progress in Polymer Science, 81*, pp. 114–143.

58. Di Noto, V., Lavina, S., Giffin, G.A., Negro, E. and Scrosati, B., 2011. Polymer electrolytes: Present, past and future. *Electrochimica Acta, 57*, pp. 4–13.

59. Meyer, W.H., 1998. Polymer electrolytes for lithium-ion batteries. *Advanced Materials, 10*(6), pp. 439–448.

60. Song, J.Y., Wang, Y.Y. and Wan, C.C., 1999. Review of gel-type polymer electrolytes for lithium-ion batteries. *Journal of Power Sources, 77*(2), pp. 183–197.

61. Chen, N., Zhang, H., Li, L., Chen, R. and Guo, S., 2018. Ionogel electrolytes for high-performance lithium batteries: A review. *Advanced Energy Materials, 8*(12), p. 1702675.

Energy Materials and Energy Harvesting

62. Croce, F., Appetecchi, G.B., Persi, L. and Scrosati, B., 1998. Nanocomposite polymer electrolytes for lithium batteries. *Nature, 394*(6692), p. 456.
63. Mehta, M.A., Fujinami, T. and Inoue, T., 1999. Boroxine ring containing polymer electrolytes. *Journal of Power Sources, 81*, pp. 724–728.
64. Sun, X., Lee, H.S., Yang, X.Q. and McBreen, J., 2002. Using a boron-based anion receptor additive to improve the thermal stability of $LiPF_6$-based electrolyte for lithium batteries. *Electrochemical and Solid-State Letters, 5*(11), pp. A248–A251.
65. Tabata, S.I., Hirakimoto, T., Tokuda, H., Susan, M.A.B.H. and Watanabe, M., 2004. Effects of novel boric acid esters on ion transport properties of lithium salts in nonaqueous electrolyte solutions and polymer electrolytes. *The Journal of Physical Chemistry B, 108*(50), pp. 19518–19526.
66. Matsumi, N., Sugai, K. and Ohno, H., 2002. Selective ion transport in organoboron polymer electrolytes bearing a mesitylboron unit. *Macromolecules, 35*(15), pp. 5731–5733.
67. Zhang, Z., Shao, Y., Lotsch, B., Hu, Y.S., Li, H., Janek, J., Nazar, L.F., Nan, C.W., Maier, J., Armand, M. and Chen, L., 2018. New horizons for inorganic solid state ion conductors. *Energy & Environmental Science, 11*(8), pp. 1945–1976.
68. Zheng, F., Kotobuki, M., Song, S., Lai, M.O. and Lu, L., 2018. Review on solid electrolytes for all-solid-state lithium-ion batteries. *Journal of Power Sources, 389*, pp. 198–213.
69. Bachman, J.C., Muy, S., Grimaud, A., Chang, H.H., Pour, N., Lux, S.F., Paschos, O., Maglia, F., Lupart, S., Lamp, P. and Giordano, L., 2016. Inorganic solid-state electrolytes for lithium batteries: Mechanisms and properties governing ion conduction. *Chemical Reviews, 116*(1), pp. 140–162.
70. Rossbach, A., Tietz, F. and Grieshammer, S., 2018. Structural and transport properties of lithium-conducting NASICON materials. *Journal of Power Sources, 391*, pp. 1–9.
71. Chen, S., Xie, D., Liu, G., Mwizerwa, J.P., Zhang, Q., Zhao, Y., Xu, X. and Yao, X., 2018. Sulfide solid electrolytes for all-solid-state lithium batteries: Structure, conductivity, stability and application. *Energy Storage Materials, 14*, pp. 58–74.
72. Kanno, R. and Murayama, M., 2001. Lithium ionic conductor thio-LISICON: The $Li_2SGeS_2P_2S_5$ system. *Journal of the Electrochemical Society, 148*(7), pp. A742–A746.
73. Zhao, Y. and Daemen, L.L., 2012. Superionic conductivity in lithium-rich antiperovskites. *Journal of the American Chemical Society, 134*(36), pp. 15042–15047.
74. Li, Y., Xu, H., Chien, P.H., Wu, N., Xin, S., Xue, L., Park, K., Hu, Y.Y. and Goodenough, J.B., 2018. A perovskite electrolyte that is stable in moist air for lithium-ion batteries. *Angewandte Chemie International Edition, 57*(28), pp. 8587–8591.
75. Xu, K. and von Cresce, A., 2011. Interfacing electrolytes with electrodes in Li ion batteries. *Journal of Materials Chemistry, 21*(27), pp. 9849–9864.
76. Zhao, Y. and Daemen, L.L., 2012. Superionic conductivity in lithium-rich antiperovskites. *Journal of the American Chemical Society, 134*(36), pp. 15042–15047.
77. Wu, J.F., Pang, W.K., Peterson, V.K., Wei, L. and Guo, X., 2017. Garnet-type fast Li-ion conductors with high ionic conductivities for all-solid-state batteries. *ACS Applied Materials & Interfaces, 9*(14), pp. 12461–12468.
78. Thangadurai, V., Narayanan, S. and Pinzaru, D., 2014. Garnet-type solid-state fast Li ion conductors for Li batteries: Critical review. *Chemical Society Reviews, 43*(13), pp. 4714–4727.
79. Montanino, M., Passerini, S. and Appetecchi, G.B., 2015. Electrolytes for rechargeable lithium batteries. In *Rechargeable Lithium Batteries* Franco A. (Ed.) Woodhead Publishing: Elsevier. London. pp. 73–116.
80. Meesala, Y., Liao, Y.K., Jena, A., Yang, N.H., Pang, W.K., Hu, S.F., Chang, H., Liu, C.E., Liao, S.C., Chen, J.M. and Guo, X., 2019. An efficient multi-doping strategy to enhance Li-ion conductivity in the garnet-type solid electrolyte $Li_7 La_3 Zr_2 O_{12}$. *Journal of Materials Chemistry A, 7*(14), pp. 8589–8601.

81. Manthiram, A., Yu, X. and Wang, S., 2017. Lithium battery chemistries enabled by solid-state electrolytes. *Nature Reviews Materials*, 2(4), pp. 1–16.
82. Ratner, M.A., Johansson, P. and Shriver, D.F., 2000. Polymer electrolytes: Ionic transport mechanisms and relaxation coupling. MRS *Bulletin*, 25(3), pp. 31–37.
83. Degner, T., Dimeas, A., Engler, A., Gil, N., de Muro, A.G., Jiménez-Estévez, G., Kariniotakis, G., Korres, G., Madureira, A., Mao, M. and Marnay, C., 2014. In *Microgrids: Architectures and Control*. Hatziargyriou N. (Ed.) Wiley-IEEE Press.
84. Karshenas, H.R., Daneshpajooh, H., Safaee, A., Jain, P. and Bakhshai, A., 2011. Bidirectional dc-dc converters for energy storage systems. In *Energy Storage in the Emerging Era of Smart Grids*. Carbone R. (Ed.) InTech. London. pp. 161–178.
85. Du, P. and Lu, N., 2014. *Energy Storage for Smart Grids: Planning and Operation for Renewable and Variable Energy Resources (VERs)*. Academic Press: Elsevier. London.
86. Dunn, B., Kamath, H. and Tarascon, J.M., 2011. Electrical energy storage for the grid: A battery of choices. *Science*, 334(6058), pp. 928–935.
87. Larcher, D. and Tarascon, J.M., 2015. Towards greener and more sustainable batteries for electrical energy storage. *Nature Chemistry*, 7(1), p. 19.
88. Yang, Z., Zhang, J., Kintner-Meyer, M.C., Lu, X., Choi, D., Lemmon, J.P. and Liu, J, 2011. Electrochemical energy storage for green grid. *Chemical Reviews*, 111(5), pp. 3577-3613.
89. Hesse, H.C., Schimpe, M., Kucevic, D. and Jossen, A., 2017. Lithium-ion battery storage for the grid—a review of stationary battery storage system design tailored for applications in modern power grids. *Energies*, 10(12), p. 2107.
90. Lennon, A., Jiang, Y., Hall, C., Lau, D., Song, N., Burr, P., Grey, C.P. and Griffith, K.J., 2019. High-rate lithium ion energy storage to facilitate increased penetration of photovoltaic systems in electricity grids. *MRS Energy & Sustainability*, 6(2), pp. 1–18.
91. Saft industrial Battery Group: DAtenblatt High Energy Lithium-Ion Module 48 V/2.2 kWh, 2010.
92. Scrosati, B. and Garche, J., 2010. Lithium batteries: Status, prospects and future. *Journal of Power Sources*, 195(9), pp. 2419–2430.
93. Ruiz, V.R., Kriston, A., Adanouj, I., Destro, M., Fontana, D. and Pfrang, A., 2018. The effect of charging and discharging lithium iron phosphate-graphite cells at different temperatures on degradation. *JoVE (Journal of Visualized Experiments)*, (137), p. e57501.
94. Samsung SDI. 2016. *Basic Specification of 94 Ah Lithium-ion Battery Cell*. Samsung SDI: Seongnam-si, Korea.
95. Panasonic. 2012. *Data Sheet NCR 18650B Lithium-Ion Battery Cell*. Panasonic: Osaka, Japan.
96. N.H.P.L.I.C.A.B. 2012. A123 Systems, Data Sheet.
97. https://www.catlbattery.com/en/.

6 Advanced Processing of Superalloys for Aerospace Industries

Swadhin Kumar Patel, Biswajit Swain, and Ajit Behera
NIT Rourkela

CONTENTS

6.1 Introduction 109
6.2 Demands and Improvements of Aircrafts 110
6.3 What Is a Superalloy? 111
6.4 Types of Superalloys and Their Important Phases 112
    6.4.1 Ni-Based Superalloys 112
    6.4.2 Fe-Ni-Based Superalloys 114
    6.4.3 Co-Based Superalloys 114
6.5 Fabrication Processes of Superalloys 115
    6.5.1 Investment Casting 115
    6.5.2 Directional Solidification 116
    6.5.3 Powder Metallurgy 116
        6.5.3.1 Conventional Press-and-Sinter 116
        6.5.3.2 Hot Isostatic Pressing (HIP) 117
        6.5.3.3 Self-Propagating High-Temperature Synthesis (SHS) 117
        6.5.3.4 Spark Plasma Sintering (SPS) 119
        6.5.3.5 Microwave Sintering 119
        6.5.3.6 Metal Injection Moulding (MIM) 119
        6.5.3.7 Additive Manufacturing 121
6.6 Conclusions 121
References 122

6.1 INTRODUCTION

The first powered aircraft made by Wilbur Wright and Orville Wright in 1903 is the first successful aircraft and those two are known as the inventors of the airplane. After several trials, they came up with that powered vehicle which took us to the aerial age. At that time, steel was the most common material for structural applications. Also in these days, steel is used for most structural building applications due to its cost effectiveness over other materials and can be handled and manufactured easily. Since Iron

110   Functional and Smart Materials

Age, we are familiar with iron and with its various manufacturing processes. However, today people do research over those conventional materials and able to overcome the problems such as high-temperature resistance, corrosion resistance, creep, fatigue, specific weight, etc.

6.2 DEMANDS AND IMPROVEMENTS OF AIRCRAFTS

As time goes on, after First World War demands of aerial combat increased. For the increase in performance, light weight and fuel efficiency was the major criteria. Communication distance, carrier for explosives (bomber), cargos and soldiers were more focused in the Second World War. As time goes on, design of the aircrafts are changed. Based on the functionality of those aircrafts size, speed varies. More fuel-efficient engines were developed over the time, and some of the generations of aerospace engines are mentioned in Table 6.1. First generation has a capability of engine service temperature from 1050°C to 1076°C. As the generation evolves, service temperature increased up to 1100°C, 1124°C, 1126°C and 1130°C, respectively. Recently developed materials such as Co-Re based and Pt-Al based can withstand service temperatures of > 1300°C. All these service temperatures are calculated for 100 h of creep rupture for 140 MPa load.

As shown in Table 6.1, superalloys are the most suitable candidate for aerospace industries. Nickel percentage is more than the other alloying elements, which makes clear of its importance due to the properties of it and the phases made by it. All these details will be discussed later.

Design of the aero engine is focused on the improvement of the fuel efficiency and increment of the weight to power ratio. Carnot equation for a heat engine is mentioned in Equation 6.1.

TABLE 6.1
Materials Used in Different Generations of Aero-Engines

Generation	Alloy	Density (gm/cc)	Major Alloying Elements						
			Ni	Co	Cr	Mo	Ti	Al	W
1st	CM 247 LC	8.5	61.7	9.2	8.1	0.5	0.7	5.6	9.5
	Rene N4	8.5	62	7.5	9.8	1.5	3.5	4.2	6
	AM 1	8.6	63.5	6.5	7.5	2	1.2	5.2	5.5
2nd	CMSX 4	8.7	61.7	9	6.5	0.6	1	5.6	6
	PW 1484	8.95	59.4	10	5	2	—	5.6	6
	ZS-32	8.8	60.7	9.3	5	2	—	6	6
	Rene N5	8.63	63.1	7.5	7	1.5	—	6.2	5
3rd	CM SX10	9.05	69.6	3	2	0.4	0.2	5.7	5
	Rene N6	8.97	57.3	12.5	4.2	1.4	—	5.8	6
	DMS 4	9.08	63	7.9	2.4	—	—	5.2	5.5
4th	MC544	8.75	70.3	<0.2	4	1	0.5	6	5
	PW1497	9.2	50.6	16.5	2	2	—	5.55	6
5th	TMS 196	9.01	59.7	5.6	4.6	2.4	—	5.6	5

Superalloys for Aerospace Industries

$$\eta = 1 - \left(T_{\text{low}} / T_{\text{high}}\right) \qquad (6.1)$$

Where, η: efficiency of engine,
T_{high}: engine temperature (source),
T_{low}: ambient temperature (sink).

To improve the efficiency of the engine the term "$T_{\text{low}}/T_{\text{high}}$" should be minimum. However, we cannot control the ambient temperature. Therefore, to decrease that value, more engine temperature is required. It means increase in service temperature of the engine material can give us better engine performance. Because of the superalloys, we got such higher attainable service temperatures. Due to high melting point, researchers are trying to develop ceramic-based and carbon-based materials for different aero engine parts. Introduction of advanced cooling system such as vents in turbine blades can increase the capacity of service temperature of the blades.

As shown in Figure 6.1, 0–2 region is the air intake part, and then it is compressed through several stages of compressors in 2–3 region. Maximum heat generation takes place in 3–4 region as here fuel burns in presence of air and then passes through turbines shown in 4–5 region. This is the place where mechanical power is achieved due to the fuel combustion, and the exhaust gas finally ejects out from this system by region 5–8. The variation of pressure inside the engine is also similar to this type of temperature variation.

6.3 WHAT IS A SUPERALLOY?

Alloys are the combinations of a metal with one or more alloying elements. They may exist as single or multi-phase solid solution. These have metallic bonds with in the metal atoms. With a defined stoichiometry and crystal structure, alloys can exist as intermetallic compounds. Due to alloying of elements, different phases and intermetallic formed in it give varieties of properties compared to the parent materials. Superalloys are particularly those alloys, which can give good strength, resistance to surface degradation, creep and fatigue at higher service temperatures [1]. Generally, >540°C (1000°F) is considered as the high service temperature environment [2]. In another way, superalloys are the alloys that can be operated at 0.7 times of their absolute melting temperature.

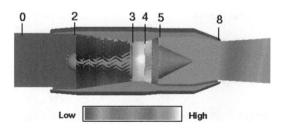

FIGURE 6.1 Temperature variation in aero-engines. (Collected from NASA Glenn Research Center's Website.)

6.4 TYPES OF SUPERALLOYS AND THEIR IMPORTANT PHASES

As shown in Figure 6.2 superalloys are generally categorized as three types. Ni based, Co based and Ni-Fe based. Other elements in the figure with elevations are the alloying elements. Heights of the elements in the figure show their weightage of choice as an alloying element. Table 6.2 provides some commonly used superalloys in the industries. Majorly it was found that all types of superalloys need Ni, which acts as a face-centred cubic (FCC) stabilizer for these alloys.

6.4.1 Ni-Based Superalloys

Major element of this category is Ni and in addition to that other alloying elements such as Ti, Al, Cr, Mo, Nb, Hf, W, Zr, B, C are also present in lesser amount. Austenitic

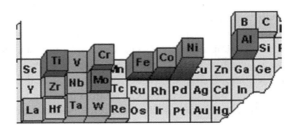

FIGURE 6.2 Alloying elements in superalloys.

TABLE 6.2
Classification of Some Wrought Superalloys [2]

	Solid-Solution Alloys	Precipitation-Hardening Alloys
Iron–nickel base	AlloyN-155 (Multimet), Haynes 556, I9-9 DL, Incoloy (800, 800HT, 800 H, 801, 802) **Fe %: (29%–66.8%)** **Ni %: (9%–33%)**	A-286, Discaloy, Incoloy (903, 907, 909, 925), Pyromet CTX-1, V-57, W-545 **Fe %: (29%–55.8%)** **Ni %: (26%–44%)**
Nickel base	Haynes (214, 230), Inconel (600, 601, 617, 625), RA333, Hastelloy (B, N, S, W, X, C-276), Haynes (HR-120, HR-160), Nimonic (75, 86) **Ni %: (37%–76.5%)**	Astroloy, Custom Age 625 PLUS, Haynes (242, 263, R-41), Inconel (100, 102, 702, 706, 718, 721, 722, 725, 751, X-750), IN-100, Incoloy901, M-252, MERL-76, Nimonic (80A, 90, 95, 100, 105, 115), C-263, C-1023, GTD 222, Pyromet (860, 31), Refractaloy 26, Rene (41, 88, 95, 100, 220), Udimet (500, 520, 630, 700, 710, 720, 720LI), Unitemp AF2-1DA, Waspaloy **Ni %: (38%–79.5%)**
Cobalt base	Haynes (25 (L605), 188), Alloy S-816, MP35-N, MP159, Stellite B, UMCo-50 **Ni %: (0%–35%)**	

Superalloys for Aerospace Industries

FCC crystal structured γ phase (solid solution of Ni_3(Al, Ti)) is the primary matrix in these alloys. Secondary phases like ordered FCC structured γ' (Ni_3(Al, Ti), body-centred tetragonal (BCT) ordered phase of Ni_3Nb (γ''), hexagonal ordered Ni_3Ti (η) phase and orthorhombic Ni_3Nb (δ) intermetallic phases are present in these superalloys [3]. Solid solution strengthening and precipitation hardening are the two major strengthening mechanisms in Ni-based superalloys (Figures 6.3 and 6.4).

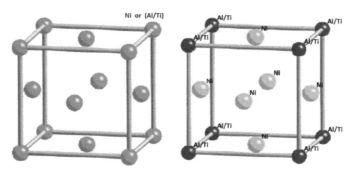

FIGURE 6.3 Crystal structure of (a) γ phase and (b) γ' phase

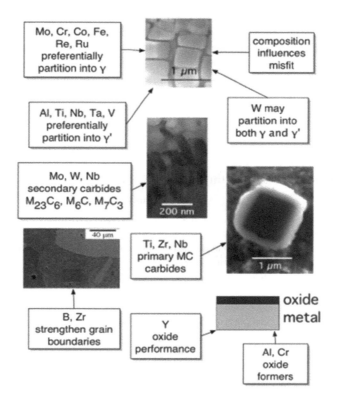

FIGURE 6.4 Role of alloying elements [4].

6.4.2 FE-NI-BASED SUPERALLOYS

Fe-based superalloys are also known as Fe-Ni-based superalloys. The major change in this type of material is to replace Ni with Fe. Lower cost of iron compared to nickel is the only explanation for the existence of such material. Addition of Cr increases the oxidation resistance. Mo, Cr help in solid solution strengthening while Ti, Al, Nb help in precipitation hardening by forming various intermetallic inside the materials. Addition of Cr and Mo reduces the solubility of Al, Ti and Nb in the matrix which helps in formation of γ' phase. So, Cr and Mo are known as primary solid solution strengthener. Precipitation hardening happens due to addition of Al and Ti by formation of intermetallics results in strengthening of the material (Table 6.3).

6.4.3 CO-BASED SUPERALLOYS

Cobalt is the primary constituent in such type of superalloys. Carbides are the principal secondary phases in cobalt-based alloys. Elements such as carbon and boron make carbides and borides. These phases are generally present at grain boundaries and helps in strengthening of the materials. They restrict the motion of grain boundaries and increase the creep resistance. Cobalt has a higher melting point than nickel but relatively lower strength. So, Co-based superalloys can be used at higher service temperatures, while their properties are less stable compared to Ni-based superalloys. In terms of resistance to thermal fatigue and weldability, Co-based superalloys are superior than Ni-based superalloys.

Other than these useful phases, topological close-packed phases (TCPs) such as P phase (orthorhombic), σ phase (tetragonal), rhombohedral R and μ phases and laves phases are also present. These are the unwanted phases for the superalloys. There

TABLE 6.3
Roles of Elements Present in the Superalloys [4]

Element(s)	Their Roles in Superalloys
Ni	Stabilizes FCC matrix, forms γ' (Ni_3(Al, Ti)), and inhibits formation of deleterious phases
Cr	Imparts oxidation and sulphidation resistance as well as solid solution strengthening, and forms grain boundary carbides
Co	Raises solvus temperature of γ' and lowers stacking fault energy (thereby making cross-slip of screw dislocations more difficult)
Mo, Ta and W	solid solution strengthening and formation of MC-type carbides
Ti	Forms γ' (Ni_3(Al, Ti)) and MC-type carbides
Al	Forms γ' (Ni_3(Al, Ti)) and improves oxidation resistance
B and Zr	Improve stress rupture properties and retard grain boundary Ni_3Ti formation
La and Y	Improve oxidation resistance
C	Formation of carbides (MC, $M_{23}C_6$ etc., type)
Nb and Ta	Form γ'' (Ni_3Nb) and MC-type carbides

Superalloys for Aerospace Industries

are two important reasons which responses in degradation of their properties. The reasons are mentioned below.

- TCPs tie up the γ and γ' phases, which results in reduction of strength in the material.
- These have needle shape structure and brittle in nature, so these are making sites for crack initiation inside the alloy (Figure 6.5).

6.5 FABRICATION PROCESSES OF SUPERALLOYS

Fabrication of these superalloys can be possible by various routes. As per the industrial importance, we can compromise the cost and time of production. In addition, accuracy of dimension is a major issue. Single crystal alloys are better in creep resistance compared to multi-crystal alloys, but inferior in terms of strength. Grain boundary plays a major role in it. Following such criteria of industrial importance, various processing routes are available and explained in the followings.

6.5.1 Investment Casting

Investment casting is an old technique to produce very intricate shapes for making turbine blades for aerospace industries (Figure 6.6). Other products such as jewelleries were used to be cast by using this process for thousands of years back. This process uses ceramic mould and wax patterns. Patterns are made by wax and dipped in ceramic slurry. After this wax inside the mould should be removed by heating. Due to this loss of wax, it is also known as lost wax casting. Before pouring the material in the mould, it is preheated for a desired temperature and molten material flows inside the mould cavity. After the completion of casting, the moulds are to be broken. Therefore, this process is a little bit costlier and labour cost is higher. On the other hand, any intricate shape with higher accuracy and surface finish can be manufactured. This process is limited to small parts, not for larger dimension parts and more time consuming compared to others.

FIGURE 6.5 Transmission electron micrograph showing (a) cuboidal γ' phases in a γ matrix and (b) M23C6 type carbide particles at the grain boundary present in diagonally from bottom left to top right [5].

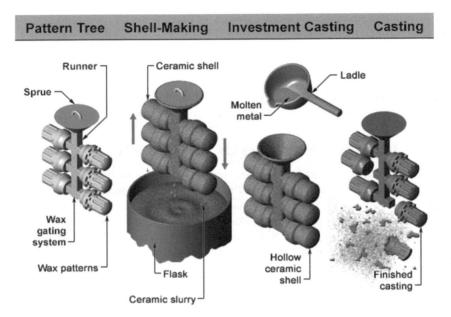

FIGURE 6.6 Investment casting procedure.

Collected from jiangsutech.com

6.5.2 Directional Solidification

As shown in Figure 6.7, this is a process which gives columnar grain due to unidirectional solidification. Molten metal is cooled from one side, and it is done by some cooling agent. In the figure, chill plates are used. The main disadvantage in casting is the formation of voids due to shrinkage of materials at varying temperatures. Mushy zone forms in between internal liquid and external solid zone. Such defects due to improper cooling can be avoided, and longitudinal grains are found in the materials as shown in the figure.

6.5.3 Powder Metallurgy

Element such as Ti is highly reactive compared to the other elements. In powder metallurgy route of fabrication, sintering is preferable in a vacuum chamber or environment of an inert atmosphere. Argon is suitable for the inert environment and used by various researchers. According to the experiment conducted by Muhammad Hussain Ismail et al., Ar is used as debinder, and sintering was done in vacuum chamber [7]. Some fabrication processes for making NiTi are briefly described.

6.5.3.1 Conventional Press-and-Sinter

Conventional press-and-sinter process is the most common powder metallurgical route for the fabrication of superalloys because of its simplicity and cost-effectiveness. In this process, powder mixture is compacted by using proper die and compressive

Superalloys for Aerospace Industries

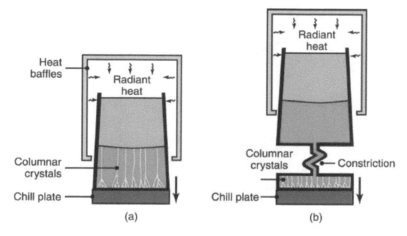

FIGURE 6.7 Schematic diagram of a directional solidification procedure [6].

loading. The sample is known as green sample. By heat treatment process, green specimen is sintered. After sintering, there is a small change in volume between the green sample and sintered sample. For biomedical application, small pores inside the material are preferable, where for high temperature structural application pores are avoidable. Porosity in the material can be calculated by using Archimedes principle as explained in the ASTM B962–08 standard [8] (Figure 6.8).

6.5.3.2 Hot Isostatic Pressing (HIP)

In this kind of fabrication process, powder mixture is encapsulated inside an evacuated chamber and gas-tight welded cans are used to transfer the gas pressure to the powder [9]. The chamber is maintained at a required high-temperature furnace by using thermocouples and surrounding furnace and the end caps are well insulated. In Figure 6.9, graphite die is used. In this process compaction and sintering happen at a single stage.

6.5.3.3 Self-Propagating High-Temperature Synthesis (SHS)

In SHS, powder mixture is kept in a quartz boat. Container enclosing the quartz boat has two pathways for gases. Suitable gas is purged through a hole and another vent is for the releasing gases from the container. By applying powder from an external electrical DC source to the powder mixture, powders are reacted with each other by ignition. This process is a time- and energy-saving process [11] (Figure 6.10).

FIGURE 6.8 Schematic diagram of a powder compaction process.

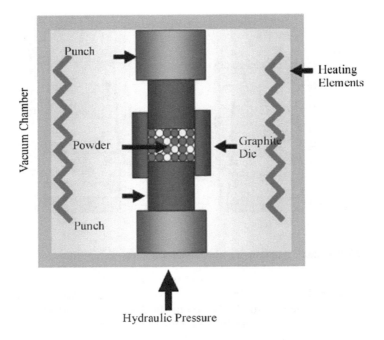

FIGURE 6.9 Schematic diagram of HIP process [10].

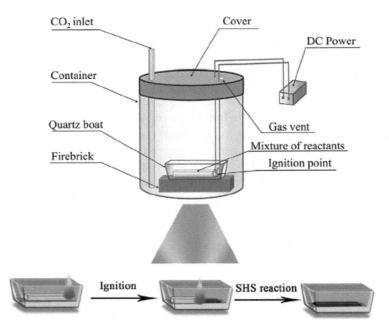

FIGURE 6.10 Schematic diagram of SHS process [12].

Superalloys for Aerospace Industries

6.5.3.4 Spark Plasma Sintering (SPS)

As explained by Ying Zhao et.al., this process consists of two punches at the two openings of the carbon die enclosed by a vacuum chamber, and punches are powered by external compressive load with pulsed DC power supply [13]. Thermocouples are used to maintain the temperature inside the die. This is a one-way sintering process. Compaction and sintering are done in a single-way process. Therefore, it takes very less time for manufacturing.

6.5.3.5 Microwave Sintering

The researcher prepared green samples by ball milling of powders and kept those samples inside a silicon carbide covered alumina sagger inside a microwave chamber filled with Ar gas [14]. Researcher used an infrared thermo detector to control the inside temperature. Here the role of SiC is to help in the heating and sintering of the sample by acting as a microwave susceptor. Major role of argon in this process is to restrict the sample to get oxidize (Figures 6.11 and 6.12).

6.5.3.6 Metal Injection Moulding (MIM)

Injection moulding can be used for both metal and polymer sample preparation from powders. In this process, powders and binder are properly mixed in a chamber and then collected in the feedstock of the equipment. From the feedstock, powder flows through a nozzle, which highly pressurize the powder and make green sample. If required, debinding of green compact is done via any heat treatment and chemical process. The product after this is known as brown product and that goes for the sintering process to give the final product (Figure 6.13).

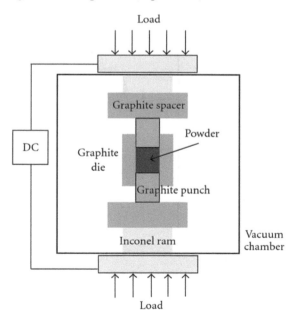

FIGURE 6.11 Schematic diagram of SPS process [15].

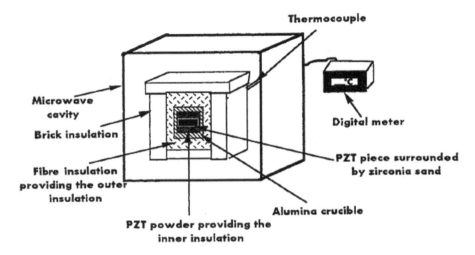

FIGURE 6.12 Schematic diagram of microwave sintering process [16].

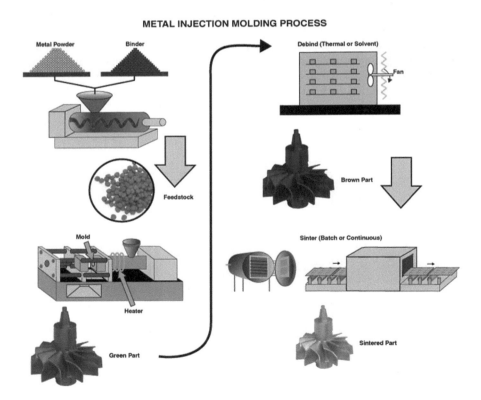

FIGURE 6.13 Schematic diagram of Injection moulding process.

Collected from jiangsutech.com

Superalloys for Aerospace Industries

6.5.3.7 Additive Manufacturing

Additive manufacturing is the technique to fabricate products using CAD model. The equipment processes the CAD model into layers of very thin slices. Those data are controlled by the machine and process the product. Powders are poured inside the fabrication chamber, and laser source is used for solidification of a single layer. This process is repeated until every layer of powder melts and fuses with each other. During this process, unwanted powders are separated from the unfinished product. To avoid any reaction with air, inert gas environment is used, though it depends on the powder we use. This is a very advanced process giving products for which no post processing is required. Accuracy is best compared to the other mentioned techniques. This process is also known as 3D printing. The major problem with this type of process is the mass production as it takes much time compared to the others. Until now, among all the processes we can say it is the best one and the quality of products is too good. The product rejection is very less in this type of manufacturing.

Expect all these types of manufacturing for the aerospace industry, thermal spraying techniques are also used to enhance the surface property of an alloy. This technique is more useful when low cost and low performance alloys for high-temperature application are used. For example, various steel products can work at higher temperatures than their normal service temperatures. Various coatings in favour of high temperature resistance can be applied over the superalloys to increase their service temperature. Coatings of ceramics or materials like carbon whose melting point is too high are the choices for such coatings. Several types of coating techniques such as plasma-spraying, oxy-fuel spraying and thermal barrier coating techniques are the most recent and hot topics for the material developers for aerospace applications. Besides this entire surface engineering, cooling techniques are also more vital for such area. If better cooling is possible, than the working temperature of a material can be increased. So, research community people are also focusing on cooling like the design of a turbine blade. The design of such equipment can guide the path of air, which can cool the components by conductive heat transfer. The recent best technique used for this is the vents in a turbine blade. As the vents are very small in size and more in numbers, they increase the surface area of heat conduction with very less degradation in mechanical properties. Laser drilling was found out to be the best process after the fabrication of a whole part. Making products with vents is very difficult as it can give uneven vent holes may cause choke during the working process and increase stress inside the material. In addition, stress concentration due to the shape of a product can be minimized due to the vents. These are the advanced processes after the fabrication of a whole product and still researches are going on to make things easier and better.

6.6 CONCLUSIONS

Nickel is the major element present in all the materials used for high-temperature applications of aerospace industries. As discussed earlier, all types of superalloys contain nickel with many alloying elements. Al and Ti are the next more important elements as they make the hard γ' phase responsible for maintaining the mechanical properties at elevated temperatures. Other elements like carbide and boride formers are helping in the strengthening of the materials. Among all the processes mentioned for the fabrication, powder technology is considered to be the best way to manufacture the products. In case of cast products, several faults can be found like void and non-homogeneity in microstructure, which leads to the low life and reduction in the quality of equipment. Heat transfer techniques are enhancing the performance of materials. Product design for better flowability of air over the surface of the products like turbine blades accompanied with drilling vents is the best solution for better heat transfer. Overall, all researchers are trying to increase the service temperature of the equipment, which is the only way to achieve better performance for high-temperature environment applications.

REFERENCES

1. R. Nasa, L. R. C. Bowman, "Superalloys: A primer and history," *9th Int. Symp. Superalloys*, vol. 3, pp. 3–6, 2000.
2. M. Kutz, *Mechanical Engineers' Handbook*, 3rd ed. Hoboken, NJ: John Wiley & Sons, Inc., 2005.
3. P. M. Mignanelli et al., "Gamma-gamma prime-gamma double prime dual-superlattice superalloys," *Scr. Mater.*, vol. 136, pp. 136–140, Jul. 2017.
4. D. V. V. Satyanarayana and N. Eswara Prasad, *Nickel-Based Superalloys*, Singapore: Springer, 2017, pp. 199–228.
5. G. S. Hillier, *Defect Energies and Deformation Mechanisms of Single Crystal Superalloys*, Cambridge: University of Cambridge, 1984.
6. D. Boruah, *Analysis of Dynamics of Bladed Discs with Monocrystalline Anisotropic Blades*, Brighton: University of Sussex, 2016.
7. M. H. Ismail, R. Goodall, H. A. Davies, and I. Todd, "Porous NiTi alloy by metal injection moulding/sintering of elemental powders: Effect of sintering temperature," *Mater. Lett.*, vol. 70, pp. 142–145, 2012.
8. G. Chen, P. Cao, and N. Edmonds, "Porous NiTi alloys produced by press-and-sinter from Ni/Ti and Ni/TiH$_2$ mixtures," *Mater. Sci. Eng. A*, vol. 582, pp. 117–125, Oct. 2013.
9. E. Schüller, O. A. Hamed, M. Bram, D. Sebold, H. P. Buchkremer, and D. Stöver, "Hot isostatic pressing (HIP) of elemental powder mixtures and prealloyed powder for NiTi shape memory parts," *Adv. Eng. Mater.*, vol. 5, no. 12, pp. 918–924, Dec. 2003.
10. S. Moustafa, W. Daoush, A. Ibrahim, and E. Neubauer, "Hot forging and hot pressing of AlSi powder compared to conventional powder metallurgy route," vol. 2, pp. 1127-1133, 2011.
11. B. Y. Li, L. J. Rong, Y. Y. Li, and V. E. Gjunter, "Synthesis of porous Ni-Ti shape-memory alloys by self-propagating high-temperature synthesis: Reaction mechanism and anisotropy in pore structure," *Acta Mater.*, vol. 48, no. 15, pp. 3895–3904, 2000.
12. X. Guo, W. Wang, Y. Yang, and Q. Tian, "Designing a large scale synthesis strategy for high quality magnetite nanocrystals on the basis of a solution behavior regulated formation mechanism," *CrystEngComm*, vol. 18, no. 47, pp. 9033–9041, Jan. 2016.
13. Y. Zhao, M. Taya, Y. Kang, and A. Kawasaki, "Compression behavior of porous NiTi shape memory alloy," *Acta Mater.*, vol. 53, no. 2, pp. 337–343, 2005.
14. C. Y. Tang, L. N. Zhang, C. T. Wong, K. C. Chan, and T. M. Yue, "Fabrication and characteristics of porous NiTi shape memory alloy synthesized by microwave sintering," *Mater. Sci. Eng. A*, vol. 528, no. 18, pp. 6006–6011, 2011.
15. G. Molénat, L. Durand, J. Galy, and A. Couret, "Temperature control in spark plasma sintering: An FEM approach," *J. Metall.*, vol. 2010, pp. 1–9, 2010.
16. P. Sharma, Z. Ounaies, V. Varadan and V. Varadan, "Dielectric and piezoelectric properties of microwave sintered PZT", Smart Materials and Structures, vol. 10, no. 5, pp. 878-883, 2001.

7 Review on Rheological Behavior of Aluminum Alloys in Semi-Solid State

R. Gupta
Research Scientist Sigma Carbon Technologies Jaipur

A. Sharma and U. Pandel
MNIT Jaipur

CONTENTS

7.1 Introduction 123
7.2 Fundamentals of Rheology 124
    7.2.1 Newtonian Fluids 124
    7.2.2 Non-Newtonian Fluids 124
7.3 Factor Affecting Viscosity 126
    7.3.1 Solid Fraction 126
    7.3.2 Temperature 126
    7.3.3 Shear Rate 128
7.4 Concluding Remarks 130
References 130

7.1 INTRODUCTION

Various characteristics of material are governed by numerous factors such as size, shape, circulation of solid particles throughout the material, solid fraction, shear rate, and viscosity of the material. During solidification, the conventionally cast alloys comprise dendritic morphology. The dendritic morphology exhibits poor ductility, feed ability, and can lead to hot tearing. The alloys being processed in a semi-solid state, this dendritic morphology must change into globular morphology. The alloy with globular morphology has better mechanical properties than the alloy having dendritic morphology [1,2].

Semi-solid metal working is a process in which alloy having partial solid and liquid behavior is injected into the die. The main requirement for semi-solid casting is a non-dendritic microstructure [3]. About 40 years ago, Spencer [4,5] remarked that stirring of alloy in semi-solid state leads to non-dendritic morphology with rheological characteristics. Modigell and Pape [6] studied rheology as an interdisciplinary science that connects physics, physical chemistry, and engineering sciences. The word "rheology" integrates the Greek words rheo denotes flow and logos implies science. Therefore, rheology deals with concurrent deformation and flow of materials [7]. The rheological behavior of material mainly reliant on viscosity of the slurry, and further the viscosity can be determined with non-dendritic morphology of SSM processing [8]. The non-dendritic morphology can be attained by several approaches such as (i) mechanical or electromagnetic stirring, (ii) thermo-mechanical treatments that involve plastic deformation monitored by heating to a semi-solid temperature, and (iii) cooling slope [9–11].

A356 aluminum alloy has been extensively studied due to its vast difference between solids and liquids temperature, excellent castability, weldability, high strength to weight ratio, and good corrosion resistance. Therefore, A356 alloy is widely used for various applications in automotive and aerospace sectors [12].

7.2 FUNDAMENTALS OF RHEOLOGY

Fundamentals of rheology are given as follow.

7.2.1 NEWTONIAN FLUIDS

Fluids in which shear stress is directly related to the shear rate are known as Newtonian fluids.

$$\tau = \eta \dot{\Upsilon} \tag{7.1}$$

Where, τ=shear stress, η=dynamic viscosity and $\dot{\Upsilon}$ =shear rate

Figure 7.1 shows a graph between shear stress and shear rate, in which the straight line is for Newtonian fluid and its slope illustrates the dynamic viscosity. Viscosity designates resistance to flow and given as internal resistance of fluids. For Newtonian fluids, the viscosity is constant irrespective of shear stress and shear rate [13]. Most gases and water are the examples of such kind of fluid [14,15].

7.2.2 NON-NEWTONIAN FLUIDS

The fluids that do not conform to Newton's law, i.e., viscosity continually varies are known as non-Newtonian fluids. The curves in Figure 7.1 for non-Newtonian fluids are non-linear and defined as:

$$\tau = K\dot{\Upsilon}^n \tag{7.2}$$

$$\eta = K\dot{\Upsilon}^{n-1} \tag{7.3}$$

Where, K= constant and n= power law index.

In Equation 7.3 for $n=1$, the viscosity is constant and shows behavior similar to Newtonian fluid. A pseudo-plastic behavior of the material is apparent for $n<1$. Various fluids such as gelatin, blood, milk are examples of fluids having pseudo-plastic

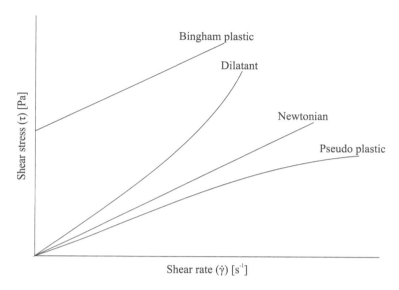

FIGURE 7.1 Different types of fluid behavior [13].

behavior. While shear thickening for $n > 1$. Sugar in water, an aqueous suspension of rice starch, etc., shows shear-thickening characteristics [16,17].

Another type of fluids in which no yield stress is observed, but for which the viscosity increases with the shear rate are acknowledged as dilatant/shear-thickening fluids and for shear thickening fluids $n > 1$ [18]. Since the viscosity of non-Newtonian fluids continually changes and mentioned as apparent viscosity [19–21].

The presence of yield stress is measured when SSM slurry is considered as Bingham fluid. Herschel–Bulkley model describes the characteristics of Bingham fluid and defined as:

$$\tau = \tau_0 + K\dot{\Upsilon}^n \qquad (7.4)$$

Where τ_0 = yield stress. Equation 7.3 is for the material that doesn't exhibit plastic behavior. If the material comprises plastic behavior, then a term related to yield stress is added and further given by Equation 7.4. This equation gives the steady-state curves and compares the theoretical and experimental results for yield stress. Various researchers [22,23] perceived extensive variations in theoretical and experimental results. For example, the shear rate exponent $n = 0.83$ for a semi-solid slurry of Sn-15Pb with a solid fraction of 0.45 [22], while $n = 1.29$ by Modigell and Koke [23]. To examine the defectiveness of the anticipated model, this anticipation was linked with the Herschel–Bulkley model (Figure 7.2).

Figure 7.2 shows that the Herschel–Bulkley model is valid for lower value of shear rate, and at higher value, it revealed variations between theoretical and experimental quantities. So, the parameters must regulate the higher and lower values demonstrated with error [24].

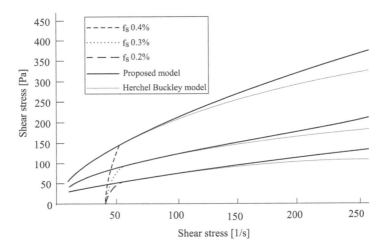

FIGURE 7.2 Comparison between proposed model and the conventional Herschel–Bulkley model [22].

7.3 FACTOR AFFECTING VISCOSITY

Different factors that affect the viscosity of slurry are described below:

7.3.1 Solid Fraction

It is an important parameter which affects the viscosity of the primary α-Al phase, e.g., dendritic α-Al phase of Al-Si alloys. In general, at constant shear rate, as solid fraction increases, the viscosity also increases, first slowly for lower value of solid fraction and rapidly for higher value solid fraction. Figure 7.3 depicts the continuous cooling curve for Sn-15Pb alloy [25].

Figure 7.3 shows that when solid fraction was < 0.6, the viscosity of Sn-15Pb alloy increases slowly and exhibits pseudo-plastic (steady-state) behavior. But for larger than 0.6 solid fraction, the viscosity varies rapidly and illustrates thixotropy (transient-state) behavior.

Lorenz et al. [26] observed the effect of an increase in solid fraction on the microstructure of Al-7Si alloys as shown in Figure 7.4. It is illustrated that as the solid fraction increases, the primary α-Al phase tends to spherical morphology having uniform grain size throughout the material.

Figure 7.4 also showed that increment in solid fraction produces coarse and round particles. At higher solid fraction, coalescence phenomena take place that lead to coarser α-Al phase and prolonged processing time support more globularization.

7.3.2 Temperature

Temperature has a substantial part in SSM processing. As temperature increases, the viscosity of slurry decreases. Temperature delivers the activation energy to the molecules of the fluid as a result the molecule starts to move on each other, i.e., comes

Rheological Behavior of Aluminum Alloys

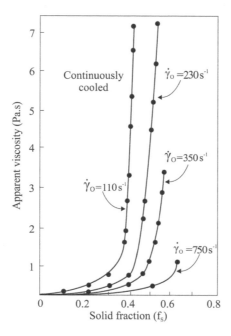

FIGURE 7.3 Apparent viscosity and solid fraction of Sn-15Pb alloy sheared at various shear rates $\dot{\gamma}_0$ [25].

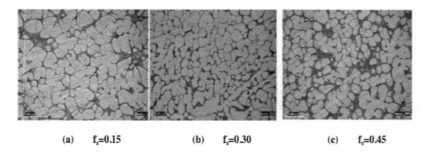

FIGURE 7.4 Optical micrographs of Al-7Si alloy water-quenched semi-solid slurry under different conditions (grain refined with SiB alloy at 1000 rpm, 5 min holding time [26]).

in the state of motion [27]. Zheng et al. [28] studied the influence of temperature on semi-solid ZL 101 (Al-7.46 Si-0.49 Mg) alloy during electromagnetic stirring.

Figure 7.5a illustrates that primary phase has rosette-like, globular-like, and particle-like morphology having coarse grains. In Figure 7.5b, most of primary phase comprises globular-like and particle-like and few shows rosette-like a-Al phase with refined grain size. Figure 7.5c depicts that primary phase has more spheroidal-like and particle-like morphology and further refined grains than grains in Figure 7.5a and b [28]. Hence, the results showed that the average grain size of primary phase decreases as pouring temperature decreases.

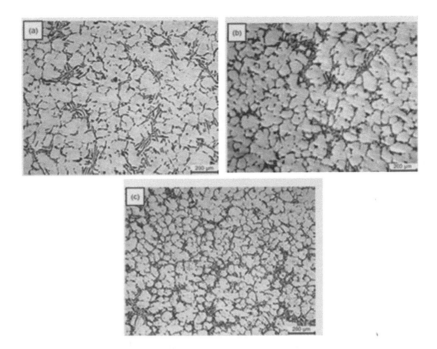

FIGURE 7.5 Morphology of α-Al phase in ZL101 alloy with stirring at various pouring temperature (a) 650, (b) 630, and (c) 615°C [28].

7.3.3 Shear Rate

More precisely the rheological characterization can be studied with isothermal steady-state experiments. The steady state, as defined earlier, is a state where the viscosity remains constant for the given value of solid fraction, shear rate, and stirring time. Hence, the steady-state viscosity is a function of solid fraction and shear rate but not of time [29]. Joly and Mehrabian [25] illustrated the pseudo-plastic behavior for Sn-15Pb alloy (Figure 7.3). Turng and Wang [30] further confirmed this kind of behavior. Figure 7.6 showed that for a fixed level of solid fraction as shear rate increases, the viscosity of the given alloy decreases and it further reached to a constant value when shear rate continually increased. This pseudo-plastic behavior was also observed by Das et al. [31].

Various investigators [32–35] also displayed the effect of shear rate on morphology of α-Al phase. They observed that with the shear rate the average grain size of primary α-Al phase decreased. Granath et al. [36] demonstrated the influence of rotation speed on the average grain size of A356 alloy formed with rapid slurry formation (RSF) process. The rotation speed varied from 600 to 1800 rpm at a given solid fraction of about 5wt.%, and the temperature of the slurry was about 627°C. The decrement in viscosity leads to de-agglomeration of α-Al phase, i.e., the features of structural breakdown in semi-solid state. Therefore, the steady-state viscosity of semi-solid state at a given shear rate can be determined by the agglomeration and de-agglomeration phenomena between the particles (Figure 7.7).

Rheological Behavior of Aluminum Alloys

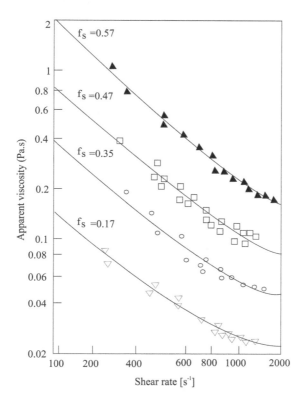

FIGURE 7.6 Steady-state apparent viscosity vs. shear rate in Sn-15Pb alloy for various solid fractions fs values [25].

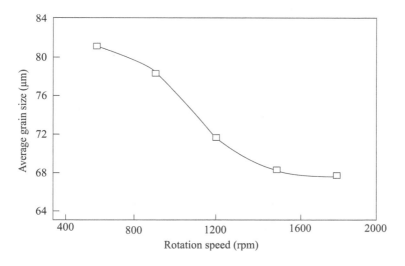

FIGURE 7.7 Effect of rotation speed on the average grain size [36].

7.4 CONCLUDING REMARKS

The consequence of various process parameters on the viscosity of SSM slurry has been studied. The major conclusion draws are as given below:

i. As the solid fraction increased, the viscosity of SSM slurry also increases; slowly at lower value of solid fraction and rapidly for higher value of solid fraction. The particles possess more globular morphology as the solid fraction increases.
ii. As the temperature increased, the viscosity of the slurry and the globular morphology decreases. But the average grain size of particles increases as the temperature increases.
iii. As the shear rate increased, the viscosity of the slurry decreases, and the particles have more globular morphology. Particle size also decreases, and the particles become fine as the shear rate increases.

REFERENCES

1. G. H. Nickodemus, C. M. Wang, M. L. Tims, J. J. Fisher, and J. J. Cardarella, Rheology of material for semi-solid metal working applications, *5th International Conference on Semi-Solid Processing of Alloy and Composites*, 1996, pp. 29–34.
2. P. Kapranos, D. H. Kirkwood, H. V. Atkinson, Development of hypereutectic Al alloys for thixoforming based on A390 composition, *6th International Conference on Semi-Solid Processing of Alloy and Composites*, 1998, pp. 341–346.
3. G. Hirt, R. Kopp, *Thixoforming Semi-Solid Metal Processing*, 1st ed. Weinheim: Wiley, Verlag GmbH, 2009, pp. 25–38.
4. D. B. Spancer, R. Mehrabian, M. C. Fleming, Patent 6, *International Materials Reviews*, 1972, pp. 1932–1935.
5. A. V. Adedayo, Effects of addition of iron filings to green moulding sand on the microstructure of grey cast iron. *Journal of Brazilian Society for Material Sciences and Engineering*, vol. 32, 2010, pp. 171–175.
6. M. Modigell, L. Paper, H. V. Atkinson, *Fundamentals of Rheology in Modeling of Semi-Solid Processing*. Aachen: Shaker Ltd, 2008, pp. 20–49.
7. O. Lashkari, R. Ghomashchi, The implication of rheology in semi-solid metal processes: An overview, *Journal of Materials Processings Technology*, vol. 182, 2007, pp. 229–240.
8. S. Ordonez, O. Bustos, R. Colas, Thermal and Microstructural Analysis of an A356 Aluminium Alloy Solidified under the Effect of Magnetic Stirring. *International Journal of Metal Casting*, vol. 3, 2009, pp. 37–41.
9. P. R. Sahm, Y. Tsutsui, M. Kiuchi, K. Ichikawa, Approaches to the production of semi-solid Al alloy in the rheocasting process, *Proceedings of the 7th International Conference on Advanced Semi-Solid processing of Alloys and Composites*, 2002, pp. 789–794.
10. B. Bobic, M. Babic, S. MItrovic, M. T. Jovanovic, I. Bobic, *Comparison of Rheological Behavior Between Semi-Solid Mixtures of Za27 Alloys and Za27-Al$_2$O$_3$ Composites*. Association of Metallurgical Engineers of Serbia, 2009.
11. H. Moller, G. Govender, W. E. Stump, P. C. Pistorious, Comparison of heat treatment response of semi-solid metal processed alloys A356 and F357, *International Journal of Cast Metals Research*, vol.23, pp.37–43, 2010.

Rheological Behavior of Aluminum Alloys

12. P. Das, K. Sudip, S. H. Chattopadhyay, P. Dutta, N. Barman, *Rheological Characterization of Semi-Solid A356 Al Alloy*. Trans. Tech Publications, Switzerland, vol. 192, 2013, pp. 329–334.

13. L. Ma, C. Denny, L. Davis, G. Obaldo, V. Gutavo, B. Canovas, *Engineering Properties of Food and Other Biological Materials: A Laboratory Manual*. Washington, DC: Washington State University, 1998, pp. 99–114.

14. N. D. Nevers, *Fluid Mechanics for Chemical Engineers*, 2nd ed., 1999, pp. 458–467.

15. T. Basner, Rehocasting of semi-solid A357 aluminium alloy, SAE Technical Paper Series, Singapore: McGraw-Hill International Editions, 2000, pp. 1–7.

16. V. Pouyafar and S.A. Sadough, An Enhanced Herschel–Bulkley Model for Thixotropic Flow Behavior of Semisolid Steel Alloys, Metallurgical and Materials Transactions B, Vol. 44, pp. 1304–1310, 2013.

17. M. C. Fleming, Behavior of metals in the semi-solid state, *Metallurgical Transactions A*, vol. 22, 1991, pp. 957c981.

18. R. A. Mclelland, N. G. Henderson, H. V. Atkinson, D. H. Kirkwood, Anomalous rheological behaviour of semi-solid alloy slurries at low shear rates, *Material Science and Engineering*, Vol. 232, pp. 110–118, 1997.

19. H. V. Atkinson, Current status of semi-solid processing of metallic materials, *Advance in Material Forming*, 2007, pp. 81–98.

20. A. Rouff, V. Favier, R. Bigot, Characterization of thixo forging steel during extrusion test, *7th International Conference on Semi-Solid Processing of Alloys and Composites*, 2002, pp. 423–428.

21. Z. Fan, J. Y. Chen, Modelling of rheological behavior of semi-solid metal slurries part 2, steady state behavior, *Journal of Material Science and Technology*, vol. 18, 2002, pp. 243–249.

22. D. H. Kirkwood, M. Suery, P. Kapranos, H. V. Atkinson, K. P. Young, M. Suery, C. H. Peng, K. K. Wang, *Semi-Solid Processing of Alloys*. Berlin, Heidelberg: Springer, vol. 124, 2010.

23. M. Modigell, J. Koke, Time dependent rheological properties of semi-solid metal alloys. *Mechanics of Time Dependent Materials*, vol. 3, 1999, pp. 15–30.

24. V. Pouya, S. A. Sadough, An enhanced Herschel-Bulkley model for thixotropic flow behavior of semi-solid steel alloys, *Metallurgical and Materials Transactions B*, vol. 44, 2013, pp. 1304–1310.

25. Y. Murakami, K. Miwa, N. Omura, S. Tada, Effects of injection speed and fraction solid on tensile strength in semi-solid injection molding of AZ91D magnesium alloy, *Materials Transactions*, vol. 53, 2012, pp. 1775–1781.

26. L. Ratke, A. Sharma, D. Kohali, Effect of process parameters on properties of Al-Si alloys cast by Rapid Slurry Formation (RSF) technique, *3rd International Conference on Advances in Solidification Processes, IOP Conference Series; Material Science and Engineering*, vol. 27, 2012, pp. 1–7.

27. T. Z. Kattamis, T. J. Piccone, Rheology of semi-solid Al-4.5Cu-1.5Mg alloy, *Material Science and Engineering A*, vol. 131, 1991, pp. 265–272.

28. L. Zheng, M. Weimin, Effect of pouring temperature on semi-solid slurry of ZL101 alloy prepared by slightly electromagnetic stirring, *China Foundry*, vol. 6, 2009, pp. 9–14.

29. Y. Fukui, D. Nara, N. Kumazawa, Evaluation of the deformation behavior of a semi-solid hypereutectic Al-Si alloy compressed in a drop-forge viscometer, *Metallurgical and Materials Transactions A*, vol. 46, 2015, pp. 1908–1916.

30. L. H. Qi, L. Z. Su, J. M. Zhou, J. T. Ghun, X. H. Hou, H. J. Li, Infiltration characteristics of liquid AZ91D alloy into short carbon fiber preform, *Journal of Alloys and Compounds*, vol. 527, 2012, pp. 10–15.

31. M. Agarwal, R. Srivastava. Influence of processing parameters on microstructure and mechanical response of a high-pressure die cast aluminum alloy. *Materials and Manufacturing Processes*, vol. 34, 2019, pp. 462–472.
32. R. Gupta, A. Sharma, U. Pandel, L. Ratke, Effect of varying shear force on microstructure of A356 alloy cast through rheo-metal process, *International Journal of Material Research*, vol. 108, pp. 648–655, 2017.
33. R. Gupta, A. Sharma, U. Pandel, L. Ratke, Wear analysis of A356 alloy cast through rheometal process, *Material Research Express*, vol. 4, 2017, pp. 1–11.
34. R. Gupta, A. Sharma, U. Pandel, L. Ratke, Effect of heat treatment on microstructures and mechanical properties of Al alloy cast through rapid slurry formation (RSF) process, *International Journal of Cast Metal Research*, vol. 30, 2017, pp. 283–292.
35. R. Gupta, A. Sharma, U. Pandel, L. Ratke, Coarsening kinetics in Al alloy cast through rapid slurry formation (RSF) process, *Acta Metallurgica Slovaca*, vol. 23, 2017, pp. 12–21.
36. C. G. Kang, J. W. Bae, B. M. Kim, The grain size control of A356 aluminum alloy by horizontal electromagnetic stirring for rheology forging, *Journal of Material Processing Technology*, Vol.187–188, pp.344–348, 2007.

8 Bio-Nanomaterials
An Inevitable Contender in Tissue Engineering

Pankaj Dipankar and Tara Chand Yadav
Indian Institute of Technology

Pallavi Saxena
Mohanlal Sukhadia University

Shanid Mohiyuddin
Indian Institute of Technology

CONTENTS

8.1 Introduction 134
8.2 Historical Aspects of Biomaterials in Tissue Engineering 136
8.3 Biomaterials: Functional Implications in Tissue Engineering 136
    8.3.1 Hydrogels 139
        8.3.1.1 Physical Hydrogels 139
        8.3.1.2 Chemical Hydrogels 140
    8.3.2 Chitosan 141
        8.3.2.1 Tissue Engineering Application 142
    8.3.3 Silk 143
        8.3.3.1 Applications of Fibroin-Based Biomaterial in
               Tissue Engineering 143
        8.3.3.2 Applications of Sericin-Based Biomaterial in
               Tissue Engineering 144
    8.3.4 Polyesters 145
        8.3.4.1 Polyhydroxyalkanoates 145
        8.3.4.2 Applications of PHAs 145
        8.3.4.3 Polyhydroxybutyrate (PHB) 147
        8.3.4.4 Application of PHB 147
    8.3.5 Calcium Phosphate Np's (CaP) 148
        8.3.5.1 CaP in Bone Regeneration 148
        8.3.5.2 CaP in Clinical Dentistry 149
    8.3.6 Hydroxyapatite 150
8.4 Summary and Conclusion 150
Acknowledgement 151
Conflict of Interest Statement 151
References 151

8.1 INTRODUCTION

The advancement in therapeutics development requires exploitation of naturally derived compounds to enable the explorations and sustainable utilization of biomaterials, overcomes the existing hurdles in clinical translation. The cost-effectiveness and the quality of product are the main rate-limiting factors in the constructive therapeutic strategy. The past few decades witnessed overwhelming appreciation in the exploration of biomaterials in biomedical applications such as tissue engineering, drug delivery systems and environmental hazard management. In the modern technological era, human beings are vulnerable to trauma, diseases and inborn defects lead to organs dysfunction (Dzobo et al. 2018). The low regenerative potential of the human body resulted in delayed self-healing or permanent damage to the tissue. The demand for organ transplant for an effected individual was increasing day by day and couldn't meet the supply (Levitt 2015). This brings an alarming necessity of tissue engineered products with enhanced regenerative potential. The application of tissue engineering involving the basics of engineering and biological principles (De Isla et al. 2010) eventually catered the tissue regeneration and found to be a game changer in the improvement of patient's ailment.

Tissue engineering approaches collaborating with nanotechnology offer extensive vistas of functional improvement in the scientific interventions. Therapeutic strategies involved with nanotechnology have been achieved a significant impact on the treatments of infectious disease (Furno et al. 2004) to cancer (Brigger, Dubernet, and Couvreur 2002). The increased surface to volume ratio of nano-formulations enables the reactiveness with ligands as functional groups. Controlled release behaviour with interchangeable diffusion rate as well as rapid absorption within the protein size influenced the vast exploitation of nano-formulations in biomedicine. Furthermore, the surface engineering ability with multiple ligands/peptides showed a greater impact in the targeted delivery systems (Xin et al. 2017). The improved pharmacokinetic property of nano-formulations entrusts the implication of the same in regenerative medicine (Paridah et al. 2016). The multidisciplinary aspects of nanotechnology by combining the different streams of physics, chemistry, biology and engineering altogether addressed the tissue engineering aspects. The scaffold-based manipulation of nanomaterials provided with better microenvironment for tissue adhesion and proliferation enables the rapid recovery (Danie Kingsley et al. 2013).

Different types of potential biomaterials can be implemented for the constructive development of multiple scaffolds for tissue engineering applications. Widely, they can be classified into three major classes such as polymers (natural and synthetic), ceramics and hydrogels. Polymer-based biomaterials such as polyester, nylon and polytetrafluoroethylene can be efficiently exploited in the surgery of soft tissues like blood vessels (Abruzzo et al. 2014), ear, nose and breast tissue (Teo et al. 2016). Furthermore, ceramics such as calcium phosphates, hydroxyapatite, calcium carbonate were effective in dental orthopaedic implants and hip replacement (Habraken et al. 2016). Additionally, the dental root implants, joint replacements, suture wires, bone plates and screws can be fulfilled by metal-based materials (gold, stainless steels, platinum, titanium and silver metals) (Sahoo et al. 1994). The composite materials such as fibre with calcium cement and multi-carbon composites were widely utilized

Bio-Nanomaterials

in joint implants, heart valves (Salernitano and Migliaresi 2003). Polymer-based tissue engineered scaffolds invite the attention of scientists due to their impactful role in the regeneration by increased biocompatibility, wide blending combination with respect to cell/tissue type, no or less immunogenicity (Yadav et al. 2019). Functional carbohydrate moieties in polysaccharides aid in the interaction with extracellular glycoproteins on the surface of the cells (Malik, Baig, and Manavalan 2019). These interactions enable the adhesion of the cell on the polysaccharide scaffolds to deliver a favourable environment for growth and proliferation of cells. The recapitulation properties of cells cannot be fully achieved by synthetic scaffold in native forms. However, many these scaffolds are fabricated by co-blending of extracellular matrix (ECM) protein and synthetic polymers (Murphy and Atala 2014). A combinatorial approach by incorporating natural and synthetic biomaterials benefited with cell-recognition sites for proper adhesion resulted in increased proliferation. Chondrocytes proliferation was estimated in elastin alone or in the combination with polyethylene glycol and polycaprolactone (Benoit et al. 2008, Smeriglio et al. 2015). Polymeric nano-scaffolds have improved functionality in comparison with metal and ceramic nano-formulations owning variable size ranges to multivariant fabrication materials (Yih and Al-Fandi 2006). Natural polymers have enhanced properties such as biodegradability, biocompatibility and less immunogenic in nature. In some synthetic polymers with high molecular weight showed accumulation of toxic products in the body upon degradation (Elzoghby, Samy, and Elgindy 2012). Furthermore, acidic by-product formation after the biodegradation of certain polymers such as poly lactide-co-glycolide (PLGA) (Keles et al. 2015) invites serious issues in real-time application. The success of the therapeutic implication greatly depends on the suitability of chosen material with less vulnerability towards detrimental effects.

Ceramic biomaterials including calcium phosphate (Denry and Kuhn 2016), calcium-based hydroxyapatite (HAP) (Cox et al. 2015) and magnesium phosphate ceramics (Kim et al. 2016a) have been widely exploited in porous scaffold manufacturing. A tuneable property in the crystallinity and chemical structure, structural similarity with hard bones and reduced hydroxyl groups (Rey et al. 2009) allows calcium phosphate-based ceramics as a front runner in bone scaffold material. This biocompatible calcium phosphate ceramics are considered as a promising candidate in bone regeneration application (Eliaz and Metoki 2017). The varying Ca/P ratio directly influences the bone regeneration as the release of calcium and phosphate ion required for calcification process (Khan et al. 2014, Dorozhkin 2013). Additionally, the potential capability of promoting osteogenesis and osseointegration by calcium phosphate bioceramic scaffolds was directly corroborated to surface chemistry (Denry and Kuhn 2016). In this perspective, a nano-bio-glass fabricated ceramic with enhanced bone regeneration (Xu et al. 2014, Stábile et al. 2016). Furthermore, the effect of ceramics and collagen type I have been investigated on mesenchymal stem cell differentiation (Pina et al. 2017).

Hydrogels are cross-linked hydrophilic polymer forming a 3D structural network, which mimics the native structure of ECM and hence frequently used for tissue engineering. The similarity in chemical properties of hydrogel with respect to tissue and graft cells by controlling the functionality of attached cells provided with excellent recovery (Tibbitt and Anseth 2009). The controllable physico-chemical property of

136 Functional and Smart Materials

hydrogel through crosslinking as well as the choice of precursor materials recalls the aptness in tissue engineering. Biodegradable hydrogels are proven platforms for the treatment of congenital heart defects and in fabricating human grafts (Mao and Mooney 2015, Tara et al. 2014). Further, different types of hydrogels fabricated through RGD peptide conjugated with various polymers such as polyethylene glycol (PEG) (Long et al. 2017), chitosan (Kim et al. 2016b) and alginate (Sun et al. 2017) showed enhanced efficacy.

Nano-scaffolds fabricated out of biomaterials showed a great potential in interacting with cells and extracellular microenvironment coping up with tissue regeneration. Such potential scaffolds act as an intermediator of signal transducer for spatial and temporal regulatory elements for tissue growth and development (Nitta and Numata 2013). In this chapter, we are comprehending different types of biomaterials used in tissue engineering applications. The history of biomaterials in different time period with respect to major contributions was summarised followed by detailed insights about biomaterials from polymers to exosome with special reference to tissue regeneration and their therapeutic implication.

8.2 HISTORICAL ASPECTS OF BIOMATERIALS IN TISSUE ENGINEERING

Over the past few decades, biomaterials have been widely used in clinical translator fields such as biomedicine and clinical dentistry. In the mid-20th century, a few efforts have been implemented in the prosthetic development in limb, filling for dental cavities, corneal implant and fracture fixatives. The regulatory authorization regarding sustainable utilization of biomaterials with respect to compatibility with human body and scientific approaches to enhance the efficacy was lacking in the initial period of biomaterials. The ancient Mayan people used nacre teeth from seashell date back in 600 AD points the biomaterial utilization even in the crude form from long back onwards. Table 8.1 shows the historic journey of biomaterials with a special emphasis on tissue engineering applications from 600 AD to 1970.

8.3 BIOMATERIALS: FUNCTIONAL IMPLICATIONS IN TISSUE ENGINEERING

Functionalised biomaterials are trustworthy candidates for tissue engineering applications such as the development of different prosthetics with multidimensional scientific applications. These materials are present abundantly as well as biocompatible in nature which suited for biomedical applications. The intrinsic structural composition of biomaterials makes them a prime choice for the transformative device fabrications. In this chapter, we concise the functional aspects of different biomaterials within the domain of tissue engineering application. The biomaterials like natural polymers such as chitosan, cellulose, silk, lignin, alginic acid, hyaluronic acid, bio-polyester (polyhydroxyalkanoates (PHAs) and polyhydroxybutyrate (PHB)) have been explained

Bio-Nanomaterials

137

TABLE 8.1
Major Landmark in the History of Materials Used in Tissue Engineering

S. No	Year	Material/ Biomaterial	Contributor	Application	References
1	600 AD	Nacre teeth from sea shell	Mayan people	Osseointegration	(Rahmati et al. 2018)
2	200 AD	Iron material	France citizens	Dental implants	(Ratner et al. 2004)
3	130–200 AD	Gold wire	Galen of Pergamon	Metallic suture for surgery	(Hossain, Roy, and Guin 2017)
4	1791	Titanium	William Gregor	Bone tissue implant	(Ananth et al. 2015)
5	1816	Lead wire	Philip Physick	Suture with no allergic response	(Abu-Faraj 2012)
6	1829	Gold, silver, lead, platinum	H. S. Levert	Biocompatibility tested in *Canis lupus familiaris*	(Stupp et al. 2005)
7	1849	Silver wire	J. Marion Sims	Metallic suture	(McGregor 1998)
8	1886	Nickel plate sheet	Carl Hansmann	Internal fixation for fractures	(Bartoníček 2010)
9	1891	Cemented ivory ball	Theodore Gluck	Hip replacement	(Gomez and Morcuende 2005)
10	1893–1912	Stainless steel	W.A. Lane	Bone fracture screws and plate	(Park and Bronzino 2002)
11	1912	Vanadium steel plates	W.D. Sherman	Medical purposes	(Greco, Prinz, and Smith 2004)
12	1925	Glass hemisphere	M. N. Smith Peterson	As a ball of hip joint	(Hernigou 2014)
13	1926	Stainless steel containing Molybdenum	M. Large	Biocompatibility	(Ratner et al. 2004)
14	1926	Molybdenum stainless steel	M.Z. Lange	Bone fracture fixation with high tensile strength	(Bhat 2002)
15	1926	Carpenter's screw	E.W. Hey-Groves	Femoral neck fracture fixation	(Bronzino 1999)
16	1931	Stainless steel	M.N. Smith-Petersen	First femoral neck fracture fixation device	(Bartonícek 2004)
17	1936	Vitallium® stainless steel	C.S. Venable, W.G. Stuck	Bone and hard tissue fixation	(Hernigou, Quiennec, and Guissou 2014)
18	1937	Polyurethane	Otto Bayer	Heart valve	(Gostev, Karpenko, and Laktionov 2018)
19	1938	Teflon (Poly tetra fluro ethylene)	Roy Plunkett	Synthetic vascular graft	(Ebert and Wolf 2011)
20	1940s	PMMA	M.J. Dorzee, A. Franceschetti	Acrylics contact lens	(Park and Bronzino 2002)
21	1942	Vitallium metal	Blakemore	Bridge for arterial defects	(Tozzi 2007)

(Continued)

TABLE 8.1 (*Continued*)
Major Landmark in the History of Materials Used in Tissue Engineering

S. No	Year	Material/ Biomaterial	Contributor	Application	References
22	1946	PMMA	R. Judet	Joint replacements	(Ramalingam et al. 2016)
23	1947	Titanium alloy	J. Kotton	Medical implants	(Wong, Peterson, Bronzino 2012)
24	1947	Polyethylene	Ingraham	Synthetic implant (mild allergic reaction)	(Mangir et al. 2019)
25	1948	Poly methyl methacrylate (PMMA)	Kevin Tuohy	Plastic Contact lens	(Chang et al. 2018)
26	1952	Cellophane	A.B. Voorhees, A. Jaretzta, A.B. Blackmore	Blood vessel replacement	(Chlupáč, Filova, and Bačáková 2009)
27	1952	Plastic	Harold Ridley	Intraocular lens (IOL)	(Trivedi et al. 2003)
28	1952	PMMA nylon ball	Charles Hufnagel	Heart valve	(Mishra 2018)
29	1953	Acrylic glue	Edward J. Haboush	Bone prosthetic fixation	(Webb and Spencer 2007)
30	1954	Silicone rubber/ silicone coated rubber grid	Mc Gregor	Artificial kidney/ dialysis membrane	(Migonney 2014)
31	1957	Polyvinyl chloride	William Kolff	Artificial heart	(McKellar 2018)
32	1958	Acrylic bone cement	J. Charnley	Hip replacement	(Hosseinzadeh et al. 2013)
33	1960	Silicone	Thomas Cronin and Frank Gerow	First breast implant	(Momoh et al. 2010)
34	1960	Silicone ball	Albert Starr	Heart valve	(Gott, Alejo, and Cameron 2003)
35	1964	Silicone tube	Folkman	Isoproterenol delivery	(Bagade et al., n.d.)
36	1967	Bio-glass	Laury Hench	Protection with nuclear radiation	(Owens et al. 2016)
37	1969	Polyurethane	Denton Cooley and William Hall	Whole Artificial heart	(Liotta 2012)
38	1970s	PEG conjugated protein	Frank Davies	In vivo enhancement of functionality	(Damodaran and Fee 2010)

with special emphasis on their important applications. Moreover, bio-derived albumin nanoparticles, liposome and exosome (biological nanoparticles) were considered for the discussion. Later, inorganic biomaterials such as hydroxyapatite and calcium phosphate Np's were reviewed for comprehensive discussion. Likewise, hydrogel and gelatin are also enlisted in the biomaterial for tissue engineering application.

Bio-Nanomaterials

8.3.1 Hydrogels

Hydrogels are extremely flexible and perceptible, hydrophilic polymeric three-dimensional composites having cross-linked architecture. They are of high absorption efficiency possess features alike mechanical durability and compositions akin to ECM. Hydrogels possess remarkable biocompatibility and tangibility, which helps in increased cellular proliferation rendering them excellent material for tissue engineering (Parhi 2017). On the basis of cross-linking, they are further categorized as physical and chemical hydrogels, respectively.

8.3.1.1 Physical Hydrogels

Physical hydrogels are formed via electrostatic/ non-covalent interaction between a cationic and anionic polymer, liked with each other through molecular entanglements and are irreversible in nature. Synthesis of physical hydrogels are devoid of any crosslinking agent, they are of low tensile strength and stability (Parhi 2017).

For chronic wounds like diabetic foot and burn wounds, zinc oxide impregnated chitosan-based hydrogels have been proved as useful bandages. These types of wound dressing bandage possess 80% porosity which allows maximum absorption of wound exudates. The prepared hydrogel showed excellent antibacterial potential along with increased cell viability, enhanced blood clotting, platelet activation with increased healing efficiency during in vitro and in vivo evaluation (Sudheesh Kumar et al. 2012). A thermo-responsive gel was prepared by blending chitosan with agarose which showed improved cytocompatibility and exchange of gases preventing wound from dehydration. Thermo-responsive behaviour was offered by agarose to the gel rendering in situ fabrication feasible while enabling them to attain appropriate shape at wound site (Miguel et al. 2014). Using freeze-thaw method, Yang et al prepared polyvinyl alcohol/chitosan hydrogel doped with lignin. However, incorporation of lignin nanoparticles led to significant decrease in swelling behaviour, but they also conferred thermal and mechanical characters as well as increased antibacterial and antioxidant potential (Yang et al. 2018). A hydrogel with first order release kinetic was prepared using an ionic cross-linker (6-phosphogluconic trisodium salt), in vitro and in vivo results showed excellent results when applied as wound dressing bandage (Martínez-Martínez et al. 2018). Fabrication of a composite hydrogel made with chitosan/heparin/poly (γ-glutamic acid) impregnated with superoxide dismutase for wound healing application manifests enhanced tensile strength with shrinkage of fibre diameter owing to heparin addition. The impregnation of superoxide dismutase enhances the healing process by promoting collagen synthesis and accumulation along with regeneration of epidermis (Zhang et al. 2017a).

Polydopamine and nanocellulose were fabricated to form hydrogel for drug delivery and wound dressing agent. Three-dimensional structural framework was offered by nanocellulose while polydopamine assisted as a photothermal mediator. Using near infra-red mediated system sustained control release of drugs may possibly be achieved (Liu et al. 2018b). Using gelatin and cellulose, hydrogel impregnated with aminated silver nanoparticles has been fabricated. The resulting product showed reasonable haemostatic property with increased antibacterial and mechanical stability.

In vivo and in vitro evaluation showed exceptional wound healing efficacy and cytocompatibility (Liu et al. 2018a). Hydrogel prepared by impregnating starch with PVA/chitosan due to the presence of hydroxyl functional groups offering exceptional cross-linking efficiency. Pore size reduction has been witnessed due to ZnO nanoparticles but simultaneously surge in antibacterial potential and enhanced mechanical stability reported (Baghaie et al. 2017). Similarly, using area and water as plasticizer, copper nanoparticles doped with starch hydrogels were prepared. Prepared hydrogel evaluation for antibacterial activity against gram-positive as well as gram-negative bacteria resulted in exceptional efficacy (Villanueva et al. 2016).

8.3.1.2 Chemical Hydrogels

Chemical hydrogels are prepared irreversible linking of polymer functional groups by covalent interactions with the help of cross-linking agents or by chemical modification. These hydrogels possess better tensile strength and are highly stable to dilapidations (Parhi 2017). ZnO and silver nanoparticles impregnated hydrogels were prepared using chitosan and PEG for wound healing application. Obtained hydrogel possesses good tensile strength, pH sensitive with improved swelling characteristics and enhanced antimicrobial efficiency against gram-positive and gram-negative bacteria in comparison to ZnO nanoparticles (control) (Liu and Kim 2012). Gelatin and carboxymethyl chitosan blend hydrogels were prepared using radiation as cross-linking mediator. Neovascularization and increased cellular proliferation were achieved while gelatin confers structural stability to the prepared hydrogel. When evaluated against cutaneous wound healing, it shows high percentage of wound closure as compared to control (Huang et al. 2013). Cross-linked hydrogels prepared with modified PVA with chitosan possessing enhanced tensile strength were reported by Zhang et al. in both swollen and dry form along with maintaining the moist environment near the wound site facilitating gaseous exchange also. Sustained drug release has been shown by this hydrogel resulting in the prevention of bacterial infections at wound site (Zhang et al. 2015). Dermal wound management has been effectively carried out using hydrogels made of fluorinated methacrylated chitosan. These bandages showed excellent transport of oxygen as compared to controls with enhancement of re-epithelialization and rapid synthesis of collagen (Patil et al. 2016).

Gelatin serves as an excellent candidate for hydrogel formation owing to its remarkable moisture-retaining efficiency and non-toxic, biodegradable and biocompatible property. However, tensile strength of gelatin-based hydrogel is low, and to overcome this drawback, cellulose impregnated nanocrystals were prepared. Gelatin-based hydrogels loaded with 15% of cellulose nanocrystals were found suitable for drug loading efficiency and controlled/sustained release of drugs to carry out further in vitro and in vivo evaluations (Ooi, Ahmad, and Amin 2016). A three-dimensional hydrogel was prepared using photo-cross-linking agent gelatin methacryloyl possess excellent tensile strength and devoid of chemical degradation along with steady firmness and elasticity which are tuneable with the dynamic exposure to light upon fabrication. The prepared hydrogel scaffold was found to observe to enhance cellular proliferation and migration, adhesion resulting in rapid tissue repair and regeneration (Zhang et al. 2017b).

Bio-Nanomaterials

Chemokine-reinforced hydrogel was prepared to foster cells towards the wound site for rapid healing of wounds. Addition of chemokine renders increased synthesis of collagen deposition, rapid re-epithelialization and neovascularization without rendering any significant change in tensile strength and swelling behaviour of hydrogel (Yoon et al. 2016).

Hydrogel formulated with gelatin-reinforced hydroxypropyl cellulose forms and interpenetrating structure using enzymatic and photo cross-linking approach. Tensile strength and optical transparency were found to be good along with excellent wound healing potential. The hydrogel prepared showed sustained release of drugs thereby acting as an antibacterial dressing bandage (Wang and Wei 2017). The reaction of the silver ions with catechol and the gelatin moieties results in the formation of hydrogels of silver nanoparticles (AgNPs). Sustained release of AgNps was observed without substantial effect on cell viability (Le Thi et al. 2018). The formation of three-dimensional sericin-impregnated methacrylic-anhydride-modified gelatin (GelMA) hydrogel scaffold was prepared to evaluate wound healing potential. The prepared hydrogel showed macroporous architecture permitting the upholding of moist and sterile environment over wound site (Chen et al. 2018).

Rapid re-epithelialization and matrix formation can be achieved by using carboxymethyl cellulose (CMC) for hydrogel formulation which aids in maintaining a moist environment near the wound site. Hydrogel formulation prepared by cross-linking hydroxyethyl cellulose with citric acid and aided with tungsten oxide helps in conferring significant anti-inflammatory and antibacterial efficacy without substantially affecting the mechanical and thermal stability of hydrogel (El Fawal et al. 2018). CMC-reinforced ZnO mesoporous silica hydrogel was formulated to alleviate wound healing. Mesoporous silica offered excellent tensile strength and swelling capacity as well as improved gaseous exchange. Sustained drug release was observed which helps to evade multiple bandage application regularly, thus ascertain as cost-effective option (Rakhshaei and Namazi 2017).

8.3.2 Chitosan

Chitosan is termed as a "wonder molecule" and the second most abundant natural biopolymer on earth, it is a linear cationic polysaccharide, mainly obtained from exo-skeleton/outer shell of insects, cell wall of fungus and crustaceans (generally shrimps and crabs). It consists of binary hetero-polysaccharide D-glucosamine (deacetylated) and N-acetyl-D-glucosamine (acetylated) units linked by β-(1–4) glycosidic bonds (Yadav et al. 2019; No and Meyers 1997). Chitosan serves as an excellent material for drug delivery, wound dressing material and tissue engineering application (3-D scaffolds) owing to the intrinsic features such as antimicrobial (Arancibia et al. 2014), antitumor (Tokoro et al. 1988; Suzuki et al. 1986), wound-healing properties (Kozen et al. 2008) (Millner et al. 2009; Ueno, Mori, and Fujinaga 2001) and haemostatic properties (Kozen et al. 2008; Millner et al. 2009). Dressing bandages/materials fabricated from chitosan-conjugated ingredients stimulate enhanced wound closure, cellular proliferation, re-epithelialization and new blood vessel formation over wound site. Chitosan-conjugated formulations reduce the possibility of amputation in affected patients (Tang et al. 2016; Devi and Dutta 2017). Tissue regeneration and remodelling involves a series of intricate processes and successive events such as blood clotting, enhanced dermal epithelialization, neovascularization (synthesis of new collagen fibres and blood vessels) and contraction of wound area (Antunes et al. 2015; Rajabian et al. 2017; Sivashankari and Prabaharan 2016; Karri et al. 2016). Chitosan reacts with components of the biological system at subcellular level and stimulates the involved process.

8.3.2.1 Tissue Engineering Application

Fabrication of polyelectrolyte complexes involving chitosan and alginate along with glycosaminoglycans (GAGs) (heparin, chondroitin sulphate and hyaluronic acid) has been prepared in form of dressing bandage (Hirano, Zhang, and Nakagawa 2001; Wang, Khor, and Lim 2001; Yan, Khor, and Lim 2000; Wang et al. 2002). Chitosan-GAG complexes were found to possess low mechanical stability but effective in releasing GAGs to assist rapid healing of wounds. Hydrogel prepared from chitosan–gelatin was observed to aid in 3D bioprinting (Ng, Yeong, and Naing 2016). Electrospun nanofibrous scaffolds reported by Sarkar et al shows good cellular proliferation, keratinocyte and fibroblast migration possessing significant mechanical stability and swelling behaviour (Sarkar et al. 2013). Salehi et al fabricated a nanofibrous mat coated with chitosan reinforced poly(L-lactic acid)/collagen. The fabricated was used for sustained release of Aloe Vera, demonstrating enhanced fibroblast attachment with better cellular proliferation and viability (Salehi et al. 2016).

Chitosan and hydroxyapatite both are highly non-immunogenic and biodegradable substances with excellent osteo-conductivity efficacy (Muzzarelli et al. 1993, Kawakami et al. 1992, Woodard et al. 2007). Hence, they are extensively employed either alone or in combination for bone tissue engineering (Dhivya et al. 2015). Chitosan possesses an outstanding efficiency to accelerate the process of bone regeneration (Ezoddini-Ardakani et al. 2012). But chitosan possesses low osteogenic activity as compared to other substitutes of bones. Addition of TGF-β (transforming growth factor beta-induced gene h3) and BMP (bone morphogenetic protein) results in enhanced bone regeneration efficiency of chitosan (Kim et al. 2002). Chitin-hydroxyapatite matrix prepared in the different ratios was employed to culture osteoblast cells and consequently implanted femur of rabbit (bone defect). The fabricated implants not only showed increased proliferation of osteoblast but also reinforced the ingrowth of adjacent tissues (Ge et al. 2004).

Chitosan functions efficiently as ECM molecule facilitating cellular proliferation and differentiation along with cell adhesion. Chitosan encompasses the capability of preserving chondrocyte cells in the undifferentiated state in addition to supporting the chondrocytes by providing cell-specific secretion (Lahiji et al. 2000). Due to this outstanding ability, chitosan is conjugated with a wide variety of polymers to form complexes for cartilage engineering. Alginate–chitosan complex developed by Iwasaki et al. enhances adhesion among chondrocyte cells (Iwasaki et al. 2004). Chitosan–alginate complex impregnated with hyaluronate and functionalized by RGD (amino acid sequence) showed significant healing in case of cartilage defect in rabbit knee within 1 month followed by complete restoration in next 6 months (Hsu et al. 2004). Likewise, chitosan-modified poly(L-lactic acid) fabricated in form of thin films showed augmented chondrocyte adhesion, proliferation and function in bovine articular chondrocyte culture (Cui et al. 2003). Chitosan microspheres impregnated with TGF-β1

Bio-Nanomaterials

scaffold manifested enhanced proliferation of chondrocyte followed by ECM production (Kim et al. 2003). Significant nerve cell growth and affinity were observed upon coating chitosan with polylysine. Nerve cell affinity was further improved by coating chitosan–polylysine with laminin and fibronectin (Haipeng et al. 2000). Laminin peptide conjugation with chitosan has been applied successfully for nerve regeneration (Matsuda et al. 2005) and neurite outgrowth (Kato et al. 2002).

8.3.3 Silk

Silk is a fibrous protein extracted from silkworm *Bombyx mori* that consists of outer and inner core protein known as sericin and fibroin, respectively. It is produced from specialized gland of epithelial cells by silkworm, flies, spiders, etc. (Preda et al. 2013). These fibrous proteins comprise different secondary structures such as α-helices, β-spirals, spacer regions and β-sheet, which are arranged from micro- to macroscopic structures (Gosline et al. 1999; Sirichaisit et al. 2003; Sheu et al. 2004; Rising et al. 2005). In the past decades, silk is used in textile industry, musical instruments string (Kundu 2014) and as suture for surgical purposes (Chellamani, Veerasubramanian, and Balaji 2014). But nowadays, it can be used as a biomaterial due to its unique property such as biocompatibility (act as a substratum for cell adhesion, proliferation, migration), stable and slow biodegradability, high tensile strength, low immunogenicity and non-cytotoxicity, high permeability for oxygen and water (Rahaiee, Zare, and Jafari 2019). Therefore, it is a promising candidate for development of biomaterial which can be used for biomedical and tissue engineering applications. The fibroin- and sericin-based biomaterials are used for different purposes which are listed below and discussed in detail.

8.3.3.1 Applications of Fibroin-Based Biomaterial in Tissue Engineering

The silk biomaterials are chemically modified by mixing with the different organic and inorganic material to produce desired material for biomedical applications (Murphy and Kaplan 2009). Different forms of silk-based matrices are formed by physical and biological modifications according to the tissue targets (Mano et al. 2007). There are several advantages of using these modified silk-based materials, including it can be easily sterilized in autoclave also by chemical and radiation treatment without affecting the property (Rnjak-Kovacina et al. 2015).

Silk-derived fibroin proteins are used as suture due to its unique property such as biocompatibility, high tensile strength and low immunogenicity (Cao and Wang 2009). Besides, fibroin-based matrices are good substratum for cell proliferation and growth for different cell types, so it has wide application in regenerative medicine and biomedical research. This biomaterial had wide applications in ligament and tendon tissue engineering and skin tissue engineering by using biomimetic scaffolds for dermal reconstruction (Alshomer, Chaves, and Kalaskar 2018), and treatment of anti-thrombogenesis (Lee et al. 1998) and hepatic tissue engineering (Diekmann, Bader, and Schmitmeier 2006) were done by using film. Additionally, bone, blood vessels engineering are done by using matrices while cartilage tissue engineering were used by hydrogel and sponge (Kim, Mauck, and Burdick 2011). Moreover, the study done by *Huang et al.* found that the silk fibroin has potential applications in axon guidance and nerve regeneration in *Rattus rattus* (Huang et al. 2012). Regenerated silk fibroin solution is blended with different natural polymers to produce nanofibres through the electrospinning process. These nanofibres are used in wound healing, drug delivery, proteins and enzyme immobilizations and other tissue engineering and biomedical applications (Huang et al. 2012).

The layering of silk fibroin protein on implant increases the biocompatibility, less immunogenicity and anticoagulant assets. Furthermore, layering of silk fibroin derived from *B. mori* reduces the microbial infection alone and with combination of silver and titanium oxide nanoparticle (Fei et al. 2013). It also improves the cell adhesion efficiency by mimicking the extracellular milieu within the coating (Hardy and Scheibel 2010). Moreover, it is also used for slow release of drugs because fibroin proteins or conjugated material with fibroin act as drug microsphere carrier responsible for controlled drug release of higher and lower molecular weight (Zhao, Li, and Xie 2015). Furthermore, researchers have made different drug delivery system by blending silk with different materials such as multi-layered films (Mandal, Kapoor, and Kundu 2009), alginate bead embedded in silk, scaffolds (Domb and Mikos 2007), microencapsulation and polyacrylamide-blended fibroin semi-interpenetrating network hydrogels (Mandal, Kapoor, and Kundu 2009).

The electronic property can be added in silk fibroin fibres by layering of silk fibre with gold nanoparticles, cadmium sulphide and carbon nanotubes (Naskar et al. 2014). While magnetic property can be imparted by coating magnetite (Dearmitt 2014). Optical property can be induced by adding green fluorescent protein with silk matrix (Putthanarat et al. 2004). The solution fibroin material is used for nanoimprinting for optical devices, lithographic patterning, microfluidic printing and microfluidic devices and also applied for identification and sensing with the help of microfabrication of meta-material structures (Bettinger et al. 2007).

8.3.3.2 Applications of Sericin-Based Biomaterial in Tissue Engineering

Since sericin is not used in textile industry, it is treated as waste so it is removed from silk fibres by a process known as degumming. In the past few years, researchers gain insight about this protein and explore their biological property so that it can modify into desired biomaterials and it can be used for biomedical applications such as cosmetics additives for enhancing fairness of body (Patil and Ferritto 2013), anticancer property, cell additive, anti-coagulant, and antioxidant, UV-resistant, anti-bacterial (Padamwar and Pawar 2004), and other applications are discussed below in details.

In the past few decades, sericin gains attention because of its unique property so it can used for different applications, such as in cosmetic industry, it is added in creams, lotion or gel for skin and hair (Barajas-Gamboa et al. 2016). It is also used in moisturizing cream for skin that has anti-aging effects (Kunz et al. 2016). Recent study reported that it has anti-tyrosinase activity, which affects the melanin production that results in the enhancement of fairness of the skin (Aramwit et al. 2018). Additionally, it is added in dietary food supplements because it has antioxidant activity which reduces the free radical formation by inhibiting lipid peroxidation process inside the body (Kunz et al. 2016).

The sericin is immunogenic in nature, and the immunogenicity depends on the amount of sericin present with the fibroin. So, sericin can be used to activate immune

Bio-Nanomaterials 145

cells for in vivo and in vitro studies (Thurber, Omenetto, and Kaplan 2015). Besides this, it has biomedicinal property like it has anti-cancer, anti-apoptotic, tumour suppression, wound-healing ability, bone regeneration (Kunz et al. 2016) and used as anticoagulant when it blended with chlorosulfonic acid (Tamada et al. 2004). Furthermore, when combined with natural or synthetic polymers, it is often used as a delivery carrier for drug, peptides, hormones, nuclear acid, etc. when it is mixed which natural or synthetic polymers (Numata and Kaplan 2010). Moreover, it is also conjugated with insulin for the treatment of diabetes mellitus because it has no immunogenic response and longer stability in blood (Zhang et al. 2006). In the past few decades, studies have shown that sericin isolated from *Antheraea mylitta* inhibits the hydrogen peroxide production and keratinocytes death upon exposure of ultraviolet radiations (Dash 2008). It has also observed that a diet supplemented with sericin reduces the tumour growth in colon in mice (Sasaki et al. 2000).

8.3.4 Polyesters

Plants and bacteria are the natural fabricator of the polymers comprising ester functional group chains, which are commonly known as polyesters. These naturally occurring polyesters are further classified under as PHAs or bacterial bioplastics. PHAs are polymers synthesized within the bacterial cells intracellularly and acts as a carbon and energy storage granules in the cell. PHAs are β-hydroxyl fatty acid biomolecules having tridecyl. It is a highly crystalline molecule with high melting point. Due to these characteristics, polyesters are considered one of the best-suited materials for biomedical purposes especially on the bone tissue engineering and drug delivery (Avérous and Pollet 2012).

8.3.4.1 Polyhydroxyalkanoates

PHAs family is composed of 3-, 4-, 5- and 6-hydroxycarboxylic acids which are known as aliphatic polyesters. PHAs synthesized via direct microbial fermentation of carbon sources in bacterial cells. PHAs are polyesters known to produce biodegradable plastics without producing any toxic waste. PHAs are polymers having more than 150 versatile monomeric units used to manufacture biodegradable plastic against the plastic made from petrochemical waste. Plastics produced via PHAs are environment friendly and recycled completely into organic waste (Park et al. 2012; Leong et al. 2014). Moreover, they are renewable and biocompatible resources over petrochemical plastics. Till now, more than 300 species have been reportedly found potent to produce polyesters which include several species of *Necator, Pseudomonas, Aeromonas, Bacillus, Alcaligenes, Enterobacter, Staphylococcus, Micrococcus* and *Rhodobacter*, etc. These bacteria comprise enzyme which cleave the ester bonds and produce water-soluble monomers in order to produce PHA granules within cytoplasm under unfavourable growth conditions.

8.3.4.2 Applications of PHAs

Other than the fabrication of plastics, PHAs also have been in use for various biomedical purposes such as drug delivery, medical implants, fabricate matrices for tissue repair, construct artificial organ, develop carriers for drug targeting, enhance the robustness of industrial microorganisms and to purify proteins (Park et al. 2012). It is also used in biofuel production, printing, photographic materials and nutritional supplements. PHAs have the potential to be blended, surface modified and/or composited with other polymers, inorganic materials and growth factors to modulate their biocompatibility, mechanical strength and degradation rates (Leong et al. 2014; Chen 2009).

Magnificent properties such as biodegradability, biocompatibility and feasibility to shape PHAs into films or porous matrices ensure the candidature of PHAs biomolecule for the application of drug delivery. It can be used as homo-polymer as well as co-polymer as short-length chain preferably for drug delivery purposes. PHAs in short length possess hydrophobic and crystalline nature which helps in releasing drug rapidly and can be degradable through surface erosion easily. Moreover, PHAs have porous surface area which helps in releasing the drug upon degradation very efficiently (Park et al. 2012; Leong et al. 2014; Chen 2009). Doxorubicin (DXR) loaded with PHAs is one of the finest examples of micro-particle of short length PHAs exhibiting high biocompatibility. However, the disadvantage of short-length PHAs is that they cannot be utilized during controlled release experimentations. For control release, medium chain length is more suitable for drug delivery purposes. Medium chain length PHAs have low melting point and crystallinity due to which release of drug in controlled manner can be targeted in contrast to the short-length chain of PHAs. PHAs having 3-hydroxyoctanoic acid and 3-hydroxyhexanoic acid carrying drugs (Tamsulosin, ketoprofen and Clonidine) as medium-length chains have shown magnificent adhesiveness against *Python reticulant's* shed skin. PHAs matrixes with all three abovementioned drugs have found efficient for transdermal drug delivery in snakes (Wang et al. 2003). However, the absorption and degradability differ in intracellular and extracellular conditions because in extracellular condition drug usually swabs off due to the blood circulation system. Therefore, nowadays to deal with this problem, PHA molecules are coupled with PEG molecules via PEGylation. Addition of PEG alters the monomeric composition and molecular mass of the PHAs which in turn alters surface chemistry of the PHA-PEG hybrid molecule and provide it amphiphilic properties (Gref et al. 1994). Similarly, control release of thymoquinone in prenatal rats has been achieved using PHA-methoxy polyethylene and glycol co-polymeric nanoparticles. targeted delivery of thymoquinone at neuronal hippocampal cells region was found to exhibit worthy bio-compatibility and very low toxicity due to the amphiphilic nature of the molecule in extracellular conditions (Shah et al. 2010). Moreover, hybrid PHAs are found as an excellent delivering agent of antibiotics. In case of highly resistant infections, they have been proved potent enough to deliver antibiotics via these hybrid polymers. In another situation, not many studies have been done yet. Wu, Wang, and Chen 2009 reported successful control release of rhodamine B isothiocyanate via PHB (Wu, Wang, and Chen 2009).

PHAs are potentially known to produce scaffolds having magnificent mechanical properties along with different inorganic phases. Other than that, PHAs are extensively in use for tissue repairing and artificial organ construction also. Properties such as cytocompatibility in addition to the checked biodegradability and non-susceptible hydrolytic degradation make PHAs suitable for tissue engineering purposes. PHAs can be easily moulded from hard to soft or brittle or vice versa as per the need and

Bio-Nanomaterials
147

scaffolds based on PHAs also support the growth of the target cell while maintaining proper supply due to which they are considered as one of the most compatible materials or tissue engineering (Masood, Yasin, and Hameed 2015). Moreover, PHAs projects diverse physiochemical nature while retaining the mechanical strength and degraded with the controlled rate while used for the tissue engineering purposes. Various bioceramics such as hydroxyapatite also have been used in order to improve the flexibility and mechanical properties of PHAs.

8.3.4.3 Polyhydroxybutyrate (PHB)

PHAs have been extensively used for tissue engineering purposes. Likewise, PHB poly 3-hydroxybutyrate which is basically either a copolymer of Poly(3-hydroxybutyrate-co-3-hydroxyvalerate) (PHBV), poly 4-hydroxybutyrate (P4HB), copolymers of 3-hydroxybutyrate and 3-hydroxyhexanoate (PHBHHx) and poly 3-hydroxyoctanoate (PHO) are among the widely used biomolecule for the tissue engineering applications (Hazer and Steinbüchel 2007; Zinn, Witholt, and Egli 2001; Hoefer 2010; Kim, Chung, and Rhee 2007). PHB is a ubiquitous, active and solvating biopolymer having low molecular weight. It is usually complexed with other biomolecules present in biological cells and presents as c-PHB in nearly all phyla (Rai et al. 2011).

8.3.4.4 Application of PHB

The study reported that PHB helps in rapid bone formation without any chronic inflammatory response after the implantation even up to 12 months (Scholz 2010). PHB usually mixed with other particulate materials such as hydroxyapatite (HA) to increase the activity and biodegradability of the composition for tissue replacement and regeneration purposes. However, amount of HA plays a vital role in attributing mechanical property and bioactivity to the composition. In a short period, formation of bonelike apatite got formed on PHB-HA hybrid at 37°C (Hazer, Lenz, and Fuller 1994). Moreover, PHBs composite materials possess same mechanical strength in compression of 62MPa which interestingly similar to the magnitude of the human bones due to this property, PHBs can be used for fracture fixation (Curley et al. 1996). In fact, a recent study reported that PHBV-HA hybrid implanted in rabbits was found morphologically, biologically and chemically compatible. Osteoblasts and osteocytes were identified throughout the interface region. The thickness of the newly formed bone significantly increased over the period of the experiment from about 130 µM at 1 month to about 770 µM at 6 months (Park, Lenz, and Goodwin 1998). Contrastingly, even being an elastic polymer PHBs are not suitable for cardiovascular tissue engineering because it degraded very rapidly when used for in vivo (Sodian et al. 2000; Hoerstrup et al. 2000). Moreover, recent study reported successful biodegradable macro-porous tubing to create a thick myocardial patch to replace myocardial infarctions (Kenar, Kose, and Hasirci 2010). Micro-fibrous material which is formed by the blend and electrospun into fibre materials of PHBV, poly (L-D, L-lactic acid) (P(L-D, L)LA) and poly(glycerol sebacate) (PGS) have been used for myocardial patching. Another study revealed axonal regeneration with low level of inflammatory infiltration (Hazari et al. 1999; Mosahebi et al. 2002). Likewise, generation of sciatic nerve in rat and spinal cord fibre regeneration after using polycaprolactone (PCL) and PGA tube is also reported (Kiyotani et al. 1995;

Novikov et al. 2002). Biodegradable patches in the gastrointestinal tract were also found compatible, and no local immune reactions were reported initially against it (Löbler et al. 2002). But after 14th day, acute inflammation is recorded by acute marker C-reactive protein levels. PHBs also help in partial replacement of bones via metallic parts and used as carriers of antibiotics to the infected bone tissues (Jagur-Grodzinski 2006). PHBs provide microporous surface structure which helps in the attachment of osteoblasts efficiently (Wang, Wu, and Chen 2004; Cool et al. 2007). PHA/PHB containing 10% and 20% HA showcases the best growth and differentiation of murine marrow osteoblasts (Doyle, Tanner, and Bonfield 1991). Cultured osteoblasts, osteoclasts and macrophages in vitro response after implantation using PHBV reported better attachment, proliferation and differentiation (Li et al. 2005). Likewise, cartilage defects also tried to repair using PHBV, and results were much better in comparison to the collagen containing calcium phosphate. PHBV shows minimal foreign-body reaction with subsequent early cartilage formation (Luklinska and Schluckwerder 2003). Recently, three-dimensional PHBHHx scaffolds found to heal the articular cartilage defect efficiently in rats (Chen and Wu 2005). The study reported that the *in vitro* chondrocytes proliferated better on the PHBHHx/PHB scaffolds than on PHB scaffolds (Zhao et al. 2003; Deng et al. 2002). Novel cartilage fabrication using collagen matrices containing calcium phosphate (CaP-Gelfix) and PHBV are promising tool for tissue engineering (Köse et al. 2005, 2004; Ye et al. 2009).

8.3.5 Calcium Phosphate NP's (CaP)

Inorganic calcium phosphate is the major constituents and building blocks of bones and teeth. In a normal human being, it was estimated that total 85% of dry weight approximated as 2 kg on an average is made of biogenic form of calcium phosphate (Dorozhkin 2011). Further, a 1–5 mM concentration of calcium phosphate in the blood circulation was reported earlier (Muddana et al. 2009). These key biological characteristics indicate the abundance and importance of the calcium phosphate in the cellular homeostasis. Such improved properties of calcium phosphate nanoparticles (CaP) can be effectively exploited in the biological system resulted with reduced side effects and improved efficacy. Apart from these biological effects, the nano-phase calcium phosphate with apposite physico-chemical properties that can progress the functionality in biological systems. High mechanical strength, improved absorbance (bioavailability), surface modification and stimuli-responsive behaviours are the main critical properties that improvise the exploration of calcium phosphate nanoparticles in multiple domain of biomedicine such as multimodal imaging device to bone tissue engineering application.

8.3.5.1 CaP in Bone Regeneration

With the close resemblance to biological bone in chemical structure and constituents, calcium phosphate nanoparticles are frequent and most common aspirant in the bone tissue engineering applications (Lu, Yu, and Chen 2018). The successful story of implementation of calcium phosphate nanoparticles in bone tissue

Bio-Nanomaterials **149**

engineering and regeneration is reporting from every part of the world. Recently, a hybrid biphasic calcium phosphate conjugated with micro-whiskers (hBCP) (Zhu et al. 2017) proved as an enhancer in bone regeneration in *in vivo* animal models. The biocompatibility of hBCP was equated via down-regulated expression pattern of inflammatory-related genes. Such improved bone regeneration by hybrid composite can be considered as a substitute to autologous bone grafts. Further, calcium phosphate cement scaffolds contained iron oxide nanoparticle (IONP-CPC) incubated with human dental pulp stem cells proved as an efficient osteogenic activator (Xia et al. 2018). This biomaterial-based tissue engineering scaffolds hold a potential reputation in cellular regeneration by a threefold increment in the ALP activity. Additionally, injectable methylcellulose hydrogel was in situ complexed with calcium phosphate nanoparticles (MC-HAP NPs) (Kim et al. 2018). A thermo-sensitive property with multiple crystalline phases has been displayed by MC-HAP NPs. The biocompatible MC-HAP NPs provided with a high bone regeneration rate when implemented in in vivo systems with increment in new mature bone formations than pristine methylcellulose hydrogel. A polymer (poly caprolactone)-templated calcium phosphate nanoparticle improves the osteogenesis in the treatment of tibial defect and promoted in bone mineralization in rats as a clinical animal (Shim et al. 2017).

8.3.5.2 CaP in Clinical Dentistry

The multimeric forms of calcium phosphate materials such as coating, cements and composite particles have been widely used in clinical dentistry. Apparently, the improved biocompatibility and compositional similarity with teeth can inevitably enhance the treatment efficacies as a material of choice. The calcium phosphate material mixed or conjugated with polymers and hydrogels potentiates the efficacy into many folds. Further, direct administration of calcium phosphate materials has been practised for remineralization with improved Ca^{2+} and $PO4^{3-}$ ion release in the disease condition. Amorphous calcium phosphate nanoparticles are widely accepted in the long run for enamel demineralization and dentin remineralization treatment. In a study, a novel amorphous calcium phosphate nanoparticles were synthesised via spray-drying yielded with a 116-nm-sized nanoparticles utilized as an implant with improved Ca^{2+} and $PO4^{3-}$ ion release in an acidic environment for caries inhibiting restorations (Xu et al. 2011). Furthermore, dentin lesions are also been treated with amorphous calcium phosphate nanoparticles and tetra calcium phosphate owning effective dentin remineralization (Weir et al. 2017). Further, enamel demineralization also attained with improved calcium and phosphate ion release at acidic pH (Melo et al. 2017). The development of a bi-layered biomaterial construct consists of PLGA (polylactide-co-glycolide acid) as polymer along with calcium phosphate for the periodontal treatment in beagle dogs (Carlo Reis et al. 2011). Chitosan hydrogel impregnated with β-tricalcium phosphate nanoparticles utilized for periapical lesion in dogs (Abdel-Fattah et al. 2017). The treatment with such composite particles resulted in new bone formation in the infected teeth post treatment confirms the testimony of the particles. A potent inhibition in the growth of *Streptococcus mutans* and immaculate dental re-mineralization by improved Ca^{2+} and HPO_4^2 ion release was accomplished by silver NPs entrapped calcium phosphate cement (Natale et al. 2017).

8.3.6 HYDROXYAPATITE

Hydroxyapatite is a class of inorganic ceramic compound having a structural and compositional similarity with hard tissues such as bone and teeth. The naturally occurring hydroxyapatite present in bone can be synthesized via various methods in large quantities proposed for multiple application platforms. Nanophase hydroxyapatite (HA) with superior surface functionalities results in the interactions with host tissue and thereby improves the tissue integration and regeneration of bone. Hydroxyapatite directly mimics the bone minerals enlist hydroxyapatite as a prominent candidate in scaffolding manufacturing as an implant in tissue engineering applications.

The intrinsic functional property of hydroxyapatite as an osteoconductive enhancer replenishes the imperative utilization in bone tissue engineering. To this regard, a microporous hydroxyapatite (HA)/β-tricalcium phosphate (β-TCP) complex nanoparticle administrated with enhanced osteo-conductivity and bone regeneration (Lee and Kim 2017). Upregulation of alkaline phosphatase and osteocalcin resulted in the differentiation of osteoblasts at HA/β-TCP (6:4) ratio. Most of the metallic implants in the tissue engineering process have been coated with hydroxyapatite which biomimicking the osteogenesis *in vivo* (Kattimani, Kondaka, and Lingamaneni 2016). Such coating improves the stiffness as well as biocompatibility of implant. Further, the redevelopment of tooth enamel was achieved by nanohydroxyapatite in a microscratched layer (Ryu et al. 2009). The damaged enamels were repaired and re-developed with improved clinical transformation. Subsequently, nanohydroxyapatite was impregnated into the PCL with matrix45S5 bio-glass (BG) used for bone regeneration (Motealleh et al. 2017). An improved compressive strength and fracture toughness in the scaffolds hold the capability of bone regeneration. The coated scaffolds showed more than 200% increment in the property.

8.4 SUMMARY AND CONCLUSION

In the dome of the 21st century, the demand for transplantation organs is increasing day by day with the direct correlation over population and industrialization. Such demands were not satisfied with organ transplantation surgeries since the reduced number of organ donors. Scientific interrogations were carried out for an alternative of such organ graft. Tissue engineered scaffolds proved as an impactful tool with better results and recuperation in diseased conditions. The success of such treatment is defined by the interaction of scaffolds with host immune, circulatory and metabolic systems. Biomaterials undeniably considered as the best choice for fabricating tissue engineered scaffolds due to its biogenic origin. The biomaterials scaffolds profoundly exhibit improved cytocompatibility and early cure from the disease. The choice of biomaterials is the key factor for the tissue regenerations and therapeutic success in bone tissue engineering, dental implants, hepatic tissue regeneration and cancer management. In this chapter, we are revealing the imperative role of biomaterials such as chitosan, silk, biopolymers such as PHAs and PHB, hydroxyapatite, calcium phosphate and hydrogel

ACKNOWLEDGEMENT

PD is thankful to the Indian Council for Medical Research (ICMR) for the fellowship. TC is grateful to the Ministry of Human Resource and Development (MHRD) for the fellowship. Sincere thanks to the University Grant Commission (UGC), Govt. of India, for providing fellowship to SM. We express our deepest thankful towards Dr. Chander Prakash for his critical reading and suggestions in our manuscript preparations.

CONFLICT OF INTEREST STATEMENT

The authors declare no potential conflicts of interest.

REFERENCES

Abdel-Fattah, W. I., S. H. El Ashry, G. W. Ali, M. A. Abdel Hamid, A. G. El-Din, and B. El-Ashry. 2017. "Regeneration of periapical lesions post-endodontic treatment and periapical surgeries in experimental animals utilizing thermo-responsive nano-beta-tricalcium phosphate/ chitosan hydrogel: A proof of concept." *Biomedical Materials (Bristol, England)* 12 (4). England: 45007. doi:10.1088/1748-605X/aa6f26.

Abruzzo, A., C. Fiorica, V. D. Palumbo, R. Altomare, G. Damiano, M. C. Gioviale, G. Tomasello, M. Licciardi, F. S. Palumbo, and G. Giammona. 2014. "Using polymeric scaffolds for vascular tissue engineering." *International Journal of Polymer Science* 2014. 1-9 http:// dx.doi.org/10.1155/2014/689390 Hindawi.

Abu-Faraj, Z. O. 2012. *Handbook of Research on Biomedical Engineering Education and Advanced Bioengineering Learning: Interdisciplinary Concepts: Interdisciplinary Concepts.* Vol. 2. IGI Global 237–243.

Alshomer, F., C. Chaves, and D. M Kalaskar. 2018. "Advances in tendon and ligament tissue engineering: Materials perspective." *Journal of Materials* 2018. Hindawi.

Ananth, H., V. Kundapur, H. S. Mohammed, M. Anand, G. S. Amarnath, and S. Mankar. 2015. "A review on biomaterials in dental implantology." *International Journal of Biomedical Science : IJBS* 11 (3). Master Publishing Group: 113–20. https://www.ncbi. nlm.nih.gov/pubmed/26508905.

Antunes, B. P., A. F. Moreira, V. M. Gaspar, and I. J. Correia. 2015. "Chitosan/arginine–chitosan polymer blends for assembly of nanofibrous membranes for wound regeneration." *Carbohydrate Polymers* 130. Elsevier: 104–12.

Aramwit, P., N. Luplertlop, T. Kanjanapruthipong, and S. Ampawong. 2018. "Effect of urea-extracted sericin on melanogenesis: Potential applications in post-inflammatory hyperpigmentation." *Biological Research* 51 (1). BioMed Central: 54. doi:10.1186/ s40659-018-0204-5.

Arancibia, M. Y, A. Alemán, M. M. Calvo, M. E. López-Caballero, P. Montero, and M. C. Gómez-Guillén. 2014. "Antimicrobial and antioxidant chitosan solutions enriched with active shrimp (Litopenaeus Vannamei) waste materials." *Food Hydrocolloids* 35. Elsevier: 710–17.

Avérous, L., and E. Pollet. 2012. "Biodegradable polymers BT." In *Environmental Silicate Nano-Biocomposites*, edited by L. Avérous and E. Pollet, 13–39. London: Springer. doi:10.1007/978-1-4471-4108-2_2.

Bagade, O., R. Dhole, P. Mane, J. Gholave, and P. Pote. n.d. "A Facet Upshot on Parenteral Ocular Implants: In Middle of Updated Perspective."

Baghaie, S., M. T. Khorasani, A. Zarrabi, and J. Moshtaghian. 2017. "Wound healing properties of PVA/starch/chitosan hydrogel membranes with nano zinc oxide as antibacterial wound dressing material." *Journal of Biomaterials Science, Polymer Edition* 28 (18). Taylor & Francis: 2220–41.

Barajas-Gamboa, J. A., A. M. Serpa-Guerra, A. Restrepo-Osorio, and C. Ã. lvarez-López. 2016. "Sericin applications: A globular silk protein." *Ingeniería y Competitividad* 18 (2). Universidad del Vall: 193–206.

Bartonícek, J. 2004. "Proximal femur fractures: The pioneer era of 1818 to 1925." *Clinical Orthopaedics and Related Research*® 419. LWW: 306–10.

Bartoníček, J. 2010. "Early history of operative treatment of fractures." *Archives of Orthopaedic and Trauma Surgery* 130 (11). Springer: 1385–96.

Benoit, D. S. W., M. P. Schwartz, A. R. Durney, and K. S. Anseth. 2008. "Small functional groups for controlled differentiation of hydrogel-encapsulated human mesenchymal stem cells." *Nature Materials* 7 (August). Nature Publishing Group: 816. doi: 10.1038/nmat2269.

Bettinger, C., K. M. Cyr, A. Matsumoto, R. Langer, J. T. Borenstein, and D. Kaplan. 2007. *Silk Fibroin Microfluidic Devices. Advanced Materials (Deerfield Beach, Fla.).* Vol. 19. doi:10.1002/adma.200602487.

Bhat, S. V. 2002. *"Biomaterials, Alpha Science International Ltd."* 13, 1–265.

Brigger, I., C. Dubernet, and P. Couvreur. 2002. "Nanoparticles in cancer therapy and diagnosis." *Advanced Drug Delivery Reviews* 54 (5). Netherlands: 631–51.

Bronzino, J. D. 1999. *Biomedical Engineering Handbook.* Vol. 2. Boca Raton, FL: CRC Press.

Cao, Y., and B. Wang. 2009. "Biodegradation of silk biomaterials." *International Journal of Molecular Sciences* 10 (4). Molecular Diversity Preservation International (MDPI): 1514–24. doi:10.3390/ijms10041514.

Carlo Reis, E. C., A. P. B. Borges, M. V. F. Araujo, V. C. Mendes, L. Guan, and J. E. Davies. 2011. "Periodontal regeneration using a bilayered PLGA/calcium phosphate construct." *Biomaterials* 32 (35). Netherlands: 9244–53. doi:10.1016/j.biomaterials.2011.08.040.

Chang, H.-C., M.-Y. Hsu, W.-T. Hsiao, and P. J.-T. Shum. 2018. "Finite element modeling of an elderly person's cornea and rigid gas permeable contact lenses for presbyopic patients." *Applied Sciences* 8 (6). Multidisciplinary Digital Publishing Institute: 855.

Chellamani, K. P., D. Veerasubramanian, and R. S. Vignesh Balaji. 2014. "Textile implants : Silk suture manufacturing technology." *Journal of Academia and Industrial Research (JAIR)* 3 (3): 127–31.

Chen, C.-S., F. Zeng, X. Xiao, Z. Wang, X.-L. Li, R.-W. Tan, W.-Q. Liu, Y.-S. Zhang, Z.-D. She, and S.-J. Li. 2018. "Three-dimensionally printed silk-sericin-based hydrogel scaffold: A promising visualized dressing material for real-time monitoring of wounds." *ACS Applied Materials & Interfaces* 10 (40). ACS Publications: 33879–90.

Chen, G.-Q. 2009. "A microbial polyhydroxyalkanoates (PHA) based bio-and materials industry." *Chemical Society Reviews* 38 (8). Royal Society of Chemistry: 2434–46.

Chen, G.-Q., and Q. Wu. 2005. "The application of polyhydroxyalkanoates as tissue engineering materials." *Biomaterials* 26 (33). Elsevier: 6565–78.

Chlupáč, J., E. Filova, and L. Bačáková. 2009. "Blood vessel replacement: 50 years of development and tissue engineering paradigms in vascular surgery." *Physiological Research* 58 (Suppl 2): S119–39.

Cool, S. M., B. Kenny, A. Wu, V. Nurcombe, M. Trau, A. I. Cassady, and L. Grøndahl. 2007. "Poly(3-hydroxybutyrate-Co-3-hydroxyvalerate) composite biomaterials for bone tissue regeneration: *In Vitro* performance assessed by osteoblast proliferation, osteoclast adhesion and resorption, and macrophage proinflammatory response." *Journal of Biomedical Materials Research Part A* 82A (3). John Wiley & Sons, Ltd: 599–610. doi:10.1002/jbm.a.31174.

Cox, S. C, J. A Thornby, G. J Gibbons, M. A Williams, and K. K Mallick. 2015. "3D printing of porous hydroxyapatite scaffolds intended for use in bone tissue engineering applications." *Materials Science and Engineering: C* 47: 237–47. doi: 10.1016/j.msec.2014.11.024.

Cui, Y. L., A. D. Qi, W. G. Liu, X. H. Wang, H. Wang, D. M. Ma, and K. D. Yao. 2003. "Biomimetic surface modification of poly (L-lactic acid) with chitosan and its effects on articular chondrocytes *in vitro*." *Biomaterials* 24 (21). Elsevier: 3859–68.

Curley, J. M, B. Hazer, R. W Lenz, and R. C. Fuller. 1996. "Production of poly (3-hydroxyalkanoates) containing aromatic substituents by *Pseudomonas oleovorans*." *Macromolecules* 29 (5). ACS Publications: 1762–66.

Damodaran, V. B., and C. Fee. 2010. "Protein PEGylation: An overview of chemistry and process considerations." *European Pharmaceutical Review* 15 (1): 18–26.

Danie Kingsley, J., S. Ranjan, N. Dasgupta, and P. Saha. 2013. "Nanotechnology for tissue engineering: Need, techniques and applications." *Journal of Pharmacy Research* 7 (2). Elsevier Ltd: 200–204. doi: 10.1016/j.jopr.2013.02.021.

Dash, R. 2008. "Molecular Characterization of Sericin from Tropical Tasar Silkworm, Antheraea Mylitta against Oxidative Stress." IIT Kharagpur.

Dearmitt, C. 2014. "Encyclopedia of polymers and composites." *Encyclopedia of Polymers and Composites*, 1–11. doi:10.1007/978-3-642-37179-0.

Deng, Y., K. Zhao, X.-f. Zhang, P. Hu, and G.-Q. Chen. 2002. "Study on the three-dimensional proliferation of rabbit articular cartilage-derived chondrocytes on polyhydroxyalkanoate scaffolds." *Biomaterials* 23 (20). Elsevier: 4049–56.

Denry, I., and L. T Kuhn. 2016. "Design and characterization of calcium phosphate ceramic scaffolds for bone tissue engineering." *Dental Materials* 32 (1): 43–53. doi: 10.1016/j.dental.2015.09.008.

Devi, N., and J. Dutta. 2017. "Preparation and characterization of chitosan-bentonite nanocomposite films for wound healing application." *International Journal of Biological Macromolecules* 104. Elsevier: 1897–1904.

Dhivya, S., S. Saravanan, T. P. Sastry, and N. Selvamurugan. 2015. "Nanohydroxyapatite-reinforced chitosan composite hydrogel for bone tissue repair *in vitro* and *in vivo*." *Journal of Nanobiotechnology* 13 (1). BioMed Central: 40.

Diekmann, S., A. Bader, and S. Schmitmeier. 2006. "Present and future developments in hepatic tissue engineering for liver support systems : State of the art and future developments of hepatic cell culture techniques for the use in liver support systems." *Cytotechnology* 50 (1–3). United States: 163–79. doi:10.1007/s10616-006-6336-4.

Domb, A., and A. G. Mikos. 2007. "Matrices and scaffolds for drug delivery in tissue engineering." *Advanced Drug Delivery Reviews*. Netherlands. doi:10.1016/j.addr.2007.05.001.

Dorozhkin, S. V. 2011. "Calcium orthophosphates: Occurrence, properties, biomineralization, pathological calcification and biomimetic applications." *Biomatter* 1 (2). Landes Bioscience: 121–64. doi:10.4161/biom.18790.

Dorozhkin, S. V. 2013. "Calcium orthophosphate-based bioceramics,". 3840–3942. doi: 10.3390/ma6093840.

Doyle, C., E. T. Tanner, and W. Bonfield. 1991. "*In vitro* and *in vivo* evaluation of polyhydroxybutyrate and of polyhydroxybutyrate reinforced with hydroxyapatite." *Biomaterials* 12 (9). Elsevier: 841–47.

Dzobo, K., N. E. Thomford, D. A. Senthebane, H. Shipanga, A. Rowe, C. Dandara, M. Pillay, and K. S. C. M. Motaung. 2018. "Advances in regenerative medicine and tissue engineering: Innovation and transformation of medicine." *Stem Cells International* 2018. doi: 10.1155/2018/2495848.

"Ebert MJ, Lyu SP, Rise MT, Wolf MF. Biomaterials for pacemakers, defibrillators and neurostimulators. Biomaterials for Artificial Organs 2011. Woodhead Publishing: 81–112.

Eliaz, N., and N. Metoki. 2017. "Calcium phosphate bioceramics: A review of their history, structure, properties, coating technologies and biomedical applications." *Materials (Basel, Switzerland)* 10 (4). MDPI: 334. doi: 10.3390/ma10040334.

Elzoghby, A. O., W. M Samy, and N. A Elgindy. 2012. "Protein-based nanocarriers as promising drug and gene delivery systems." *Journal of Controlled Release : Official Journal of the Controlled Release Society* 161 (1). Netherlands: 38–49. doi:10.1016/j.jconrel.2012.04.036.

Ezoddini-Ardakani, F., A. Navabazam, F. Fatehi, M. Danesh-Ardekani, S. Khadem, and G. Rouhi. 2012. "Histologic evaluation of chitosan as an accelerator of bone regeneration in microdrilled rat tibias." *Dental Research Journal* 9 (6). Wolters Kluwer--Medknow Publications: 694.

Fawal, G. F. E., M. M. Abu-Serie, M. A. Hassan, and M. S. Elnouby. 2018. "Hydroxyethyl cellulose hydrogel for wound dressing: Fabrication, characterization and *in vitro* evaluation." *International Journal of Biological Macromolecules* 111. Elsevier: 649–59.

Fei, X., M. Jia, X. Du, Y. Yang, R. Zhang, Z. Shao, X. Zhao, and X. Chen. 2013. "Green synthesis of silk fibroin-silver nanoparticle composites with effective antibacterial and biofilm-disrupting properties." *Biomacromolecules* 14 (12). ACS Publications: 4483–88.

Fritzsche, K., R. W. Lenz, and R. C. Fuller. 1990. "Production of unsaturated polyesters by *Pseudomonas oleovorans.*" *International Journal of Biological Macromolecules* 12 (2). Elsevier: 85–91.

Furno, F., K. S Morley, B. Wong, B. L. Sharp, P. L. Arnold, S. M. Howdle, R. Bayston, P. D. Brown, P. D. Winship, and H. J. Reid. 2004. "Silver nanoparticles and polymeric medical devices: A new approach to prevention of infection?" *The Journal of Antimicrobial Chemotherapy* 54 (6). England: 1019–24. doi:10.1093/jac/dkh478.

Ge, Z., S. Baguenard, L. Y. Lim, A. Wee, and E. Khor. 2004. "Hydroxyapatite–chitin materials as potential tissue engineered bone substitutes." *Biomaterials* 25 (6). Elsevier: 1049–58.

Gomez, P. F., and J. A. Morcuende. 2005. "Early attempts at hip arthroplasty: 1700s to 1950s." *The Iowa Orthopaedic Journal* 25. University of Iowa: 25.

Gosline, J. M., P. A. Guerette, C. S. Ortlepp, and K. N. Savage. 1999. "The mechanical design of spider silks: From fibroin sequence to mechanical function." *The Journal of Experimental Biology* 202 (Pt 23). England: 3295–3303.

Gostev, A., A. Karpenko, and P. Laktionov. 2018. Polyurethanes in cardiovascular prosthetics. *Polymer Bulletin*. doi: 10.1007/s00289-017-2266-x.

Gott, V. L, D. E. Alejo, and D. E. Cameron. 2003. "Mechanical heart valves: 50 years of evolution." *The Annals of Thoracic Surgery* 76 (6). Elsevier: S2230–39.

Greco, R. S., F. B. Prinz, and R. L. Smith. 2004. *Nanoscale Technology in Biological Systems*. Boca Raton, FL: CRC Press.

Gref, R., Y. Minamitake, M. T. Peracchia, V. Trubetskoy, V. Torchilin, and R. Langer. 1994. "Biodegradable long-circulating polymeric nanospheres." *Science* 263 (5153). American Association for the Advancement of Science: 1600–1603.

Habraken, W., P. Habibovic, M. Epple, and M. Bohner. 2016. "Calcium phosphates in biomedical applications: Materials for the future?" *Materials Today* 19 (2). Elsevier: 69–87.

Haipeng, G., Z. Yinghui, L. Jianchun, G. Yandao, Z. Nanming, and Z. Xiufang. 2000. "Studies on Nerve Cell Affinity of Chitosan-derived Materials." *Journal of Biomedical Materials Research: An Official Journal of The Society for Biomaterials, The Japanese Society for Biomaterials, and The Australian Society for Biomaterials and the Korean Society for Biomaterials* 52 (2). Wiley Online Library: 285–95.

Hardy, J. G., and T. R. Scheibel. 2010. "Composite materials based on silk proteins." *Progress in Polymer Science* 35 (9). Elsevier: 1093–1115.

Hazari, A., G. Johansson-Rudén, K. Junemo-Bostrom, C. Ljungberg, G. Terenghi, C. Green, and M. Wiberg. 1999. "A new resorbable wrap-around implant as an alternative nerve repair technique." *Journal of Hand Surgery* 24 (3). SAGE Publications: 291–95. doi: 10.1054/JHSB.1998.0001.

Bio-Nanomaterials

Hazer, B., and A. Steinbüchel. 2007. "Increased diversification of polyhydroxyalkanoates by modification reactions for industrial and medical applications." *Applied Microbiology and Biotechnology* 74 (1): 1–12. doi: 10.1007/s00253-006-0732-8.

Hazer, B., R. W. Lenz, and R. C. Fuller. 1994. "Biosynthesis of methyl-branched poly (. beta.-hydroxyalkanoate) s by *Pseudomonas oleovorans.*" *Macromolecules* 27 (1). ACS Publications: 45–49.

Hernigou, P. 2014. "Smith-Petersen and early development of hip arthroplasty." *International Orthopaedics* 38 (1). Springer Berlin Heidelberg: 193–98. doi: 10.1007/s00264-013-2080-5.

Hernigou, P., S. Quiennec, and I. Guissou. 2014. "Hip hemiarthroplasty: From Venable and Bohlman to Moore and Thompson." *International Orthopaedics* 38 (3). Springer: 655–61.

Hirano, S., M. Zhang, and M. Nakagawa. 2001. "Release of glycosaminoglycans in physiological saline and water by wet-spun chitin–acid glycosaminoglycan fibers." *Journal of Biomedical Materials Research: An Official Journal of The Society for Biomaterials, The Japanese Society for Biomaterials, and The Australian Society for Biomaterials and the Korean Society for Biomaterials* 56 (4). Wiley Online Library: 556–61.

Hoefer, P. 2010. "Activation of polyhydroxyalkanoates: Functionalization and modification." *Frontiers in Bioscience* 15: 93–121.

Hoerstrup, S. P., R. Sodian, S. Daebritz, J. Wang, E. A Bacha, D. P. Martin, A. M Moran, K. J. Guleserian, J. S. Sperling, and S. Kaushal. 2000. "Functional living trileaflet heart valves grown *in vitro.*" *Circulation* 102 (suppl_3). Am Heart Assoc: Iii–44.

Hossain, A., S. Roy, and P. S. Guin. 2017. "The importance of advance biomaterials in modern technology: A review." *Asian Journal of Research in Chemistry* 10 (4). A&V Publications: 441–53.

Hosseinzadeh HR, Emami M, Lahiji F, Shahi AS, Masoudi A, Emami S. The acrylic bone cement in arthroplasty. Arthroplasty—update (20). intechopen:101–28.

Hsu, S.-h., S. W. Whu, S.-C. Hsieh, C.-L. Tsai, D. C. Chen, and T.-S. Tan. 2004. "Evaluation of chitosan-alginate-hyaluronate complexes modified by an RGD-containing protein as tissue-engineering scaffolds for cartilage regeneration." *Artificial Organs* 28 (8). Wiley Online Library: 693–703.

Huang, W., R. Begum, T. Barber, V. Ibba, N. C. H. Tee, M. Hussain, M. Arastoo, Q. Yang, L. G. Robson, and S. Lesage. 2012. "Regenerative potential of silk conduits in repair of peripheral nerve injury in adult rats." *Biomaterials* 33 (1). Elsevier: 59–71.

Huang, X., Y. Zhang, X. Zhang, L. Xu, X. Chen, and S. Wei. 2013. "Influence of radiation crosslinked carboxymethyl-chitosan/gelatin hydrogel on cutaneous wound healing." *Materials Science and Engineering: C* 33 (8). Elsevier: 4816–24.

Isla, N. De, C. Huselstein, N. Jessel, A. Pinzano, V. Decot, J. Magdalou, D. Bensoussan, and J. F. Stoltz. 2010. "Introduction to tissue engineering and application for cartilage engineering." *Bio-Medical Materials and Engineering* 20 (3–4): 127–33. doi:10.3233/BME-2010-0624.

Iwasaki, N., S.-T. Yamane, T. Majima, Y. Kasahara, A. Minami, K. Harada, S. Nonaka, N. Maekawa, H. Tamura, and S. Tokura. 2004. "Feasibility of polysaccharide hybrid materials for scaffolds in cartilage tissue engineering: Evaluation of chondrocyte adhesion to polyion complex fibers prepared from alginate and chitosan." *Biomacromolecules* 5 (3). ACS Publications: 828–33.

Jagur-Grodzinski, J. 2006. "Polymers for tissue engineering, medical devices, and regenerative medicine. concise general review of recent studies." *Polymers for Advanced Technologies* 17 (6). Wiley Online Library: 395–418.

Karri, V. V. S. Reddy, G. Kuppusamy, S. V. Talluri, S. S. Mannemala, R. Kollipara, A. D. Wadhwani, S. Mulukutla, K. R. S. Raju, and R. Malayandi. 2016. "Curcumin loaded chitosan nanoparticles impregnated into collagen-alginate scaffolds for diabetic wound healing." *International Journal of Biological Macromolecules* 93. Elsevier: 1519–29.

Kato, K., A. Utani, N. Suzuki, M. Mochizuki, M. Yamada, N. Nishi, H. Matsuura, H. Shinkai, and M. Nomizu. 2002. "Identification of neurite outgrowth promoting sites on the laminin A3 chain G domain." *Biochemistry* 41 (35). ACS Publications: 10747–53.

Kattimani, V. S., S. Kondaka, and K. P. Lingamaneni. 2016. "Hydroxyapatite–past, present, and future in bone regeneration." *Bone and Tissue Regeneration Insights* 7. SAGE Publications Sage UK: London, England: BTRI-S36138.

Kawakami, T., M. Antoh, H. Hasegawa, T. Yamagishi, M. Ito, and S. Eda. 1992. "Experimental study on osteoconductive properties of a chitosan-bonded hydroxyapatite self-hardening paste." *Biomaterials* 13 (11). Elsevier: 759–63.

Keles, H., A. Naylor, F. Clegg, and C. Sammon. 2015. "Investigation of factors influencing the hydrolytic degradation of single PLGA microparticles." *Polymer Degradation and Stability* 119: 228–41. doi: 10.1016/j.polymdegradstab.2015.04.025.

Kenar, H., G. T. Kose, and V. Hasirci. 2010. "Design of a 3D aligned myocardial tissue construct from biodegradable polyesters." *Journal of Materials Science: Materials in Medicine* 21 (3). Springer: 989–97.

Khan, A. F., M. Saleem, A. Afzal, A. Ali, A. Khan, and A. R. Khan. 2014. "Bioactive behavior of silicon substituted calcium phosphate based bioceramics for bone regeneration." *Materials Science and Engineering: C* 35: 245–52. doi: 10.1016/j.msec.2013.11.013.

Kim, H. W., M. G. Chung, and Y. H. Rhee. 2007. "Biosynthesis, modification, and biodegradation of bacterial medium-chain-length polyhydroxyalkanoates." *The Journal of Microbiology* 45 (2): 87–97.

Kim, I. L., R. L. Mauck, and J. A. Burdick. 2011. "Hydrogel design for cartilage tissue engineering: A case study with hyaluronic acid." *Biomaterials* 32 (34): 8771–82. doi:10.1016/j.biomaterials.2011.08.073.

Kim, I.-S., J. W. Park, I. C. Kwon, B. S. Baik, and B. C. Cho. 2002. "Role of BMP, betaig-H3, and chitosan in early bony consolidation in distraction osteogenesis in a dog model." *Plastic and Reconstructive Surgery* 109 (6): 1966–77.

Kim, J.-A., J. Lim, R. Naren, H.-s. Yun, and E. K. Park. 2016a. "Effect of the biodegradation rate controlled by pore structures in magnesium phosphate ceramic scaffolds on bone tissue regeneration *in vivo*." *Acta Biomaterialia* 44: 155–67. doi: 10.1016/j.actbio.2016.08.039.

Kim, M. H., B. S. Kim, H. Park, J. Lee, and W. H. Park. 2018. "Injectable methylcellulose hydrogel containing calcium phosphate nanoparticles for bone regeneration." *International Journal of Biological Macromolecules* 109. Elsevier B.V.: 57–64. doi:10.1016/j.ijbiomac.2017.12.068.

Kim, S., Z.-K. Cui, J. Fan, A. Fartash, T. L. Aghaloo, and M. Lee. 2016b. "Photocrosslinkable chitosan hydrogels functionalized with the RGD peptide and phosphoserine to enhance osteogenesis." *Journal of Materials Chemistry. B* 4 (31): 5289–98. doi: 10.1039/C6TB01154C.

Kim, S. E., J. H. Park, Y. W. Cho, H. Chung, S. Y. Jeong, E. B. Lee, and I. C. Kwon. 2003. "Porous chitosan scaffold containing microspheres loaded with transforming growth factor-B1: Implications for cartilage tissue engineering." *Journal of Controlled Release* 91 (3). Elsevier: 365–74.

Kiyotani, T., T. Nakamura, Y. Shimizu, and K. Endo. 1995. "Experimental study of nerve regeneration in a biodegradable tube made from collagen and polyglycolic acid." *ASAIO Journal (American Society for Artificial Internal Organs: 1992)* 41 (3): M657–61.

Köse, G. T., F. Korkusuz, P. Korkusuz, and V. Hasirci. 2004. "*In vivo* tissue engineering of bone using poly (3-hydroxybutyric acid-Co-3-hydroxyvaleric acid) and collagen scaffolds." *Tissue Engineering* 10 (7–8). Mary Ann Liebert, Inc. 2 Madison Avenue Larchmont, NY 10538 USA: 1234–50.

Köse, G. T., F. Korkusuz, A. Özkul, Y. Soysal, T. Özdemir, C. Yildiz, and V. Hasirci. 2005. "Tissue engineered cartilage on collagen and PHBV matrices." *Biomaterials* 26 (25). Elsevier: 5187–97.

Bio-Nanomaterials

Kozen, B. G., S. J. Kircher, J. Henao, F. S. Godinez, and A. S. Johnson. 2008. "An alternative hemostatic dressing: Comparison of CELOX, HemCon, and QuikClot." *Academic Emergency Medicine* 15 (1). Wiley Online Library: 74–81.

Kundu, S. 2014. *Silk Biomaterials for Tissue Engineering and Regenerative Medicine.* Elsevier.

Kunz, R. I., R. M. C. Brancalhão, L. de F. C. Ribeiro, and M. R. M. Natali. 2016. "Silkworm sericin: Properties and biomedical applications." *BioMed Research International* 2016. Hindawi.

Lahiji, A., A. Sohrabi, D. S. Hungerford, and C. G. Frondoza. 2000. "Chitosan supports the expression of extracellular matrix proteins in human osteoblasts and chondrocytes." *Journal of Biomedical Materials Research* 51 (4). Wiley Online Library: 586–95.

Lee, K. Y., S. J. Kong, W. H. Park, W. S. Ha, and I. C. Kwon. 1998. "Effect of surface properties on the antithrombogenicity of silk fibroin/S-carboxymethyl kerateine blend films." *Journal of Biomaterials Science, Polymer Edition* 9 (9). Taylor & Francis: 905–14.

Lee, Y. I., and Y. J. Kim. 2017. "Effect of different compositions on characteristics and osteoblastic activity of microporous biphasic calcium phosphate bioceramics." *Materials Technology* 32 (8). Taylor & Francis: 496–504. doi:10.1080/10667857.2017.1286554.

Leong, Y. K., P. L. Show, C. W. Ooi, T. C. Ling, and J. C.-W. Lan. 2014. "Current trends in polyhydroxyalkanoates (PHAs) biosynthesis: Insights from the recombinant *Escherichia Coli.*" *Journal of Biotechnology* 180. Elsevier: 52–65.

Levitt, M. 2015. "Could the organ shortage ever be met?" *Life Sciences, Society and Policy* 11 (1). Life Sciences, Society and Policy: 1–6. doi:10.1186/s40504-015-0023-1.

Li, J., H. Yun, Y. Gong, N. Zhao, and X. Zhang. 2005. "Effects of surface modification of poly (3-hydroxybutyrate-co-3-hydroxyhexanoate)(PHBHHx) on physicochemical properties and on interactions with MC3T3-E1 cells." *Journal of Biomedical Materials Research Part A: An Official Journal of The Society for Biomaterials, The Japanese Society for Biomaterials, and The Australian Society for Biomaterials and the Korean Society for Biomaterials* 75 (4). Wiley Online Library: 985–98.

Liotta, D. S. M. E. DeBakey and D. A. Cooley. 2012. "Mike, the master assembler; Denton, the courageous fighter: A personal overview unforgettable past remembrances in the 1960s." *Open Journal of Thoracic Surgery* 02. 2012 (2): scientific research 37–45 doi:10.4236/ojts.2012.23010.

Liu, R., L. Dai, C. Si, and Z. Zeng. 2018a. "Antibacterial and hemostatic hydrogel via nanocomposite from cellulose nanofibers." *Carbohydrate Polymers* 195. Elsevier: 63–70.

Liu, Y., and H.-Il Kim. 2012. "Characterization and antibacterial properties of genipincrosslinked chitosan/poly (ethylene glycol)/ZnO/Ag nanocomposites." *Carbohydrate Polymers* 89 (1). Elsevier: 111–16.

Liu, Y., Y. Sui, C. Liu, C. Liu, M. Wu, B. Li, and Y. Li. 2018b. "A physically crosslinked polydopamine/nanocellulose hydrogel as potential versatile vehicles for drug delivery and wound healing." *Carbohydrate Polymers* 188. Elsevier: 27–36.

Löbler, M., M. Saß, C. Kunze, K.-P. Schmitz, and U. T. Hopt. 2002. "Biomaterial implants induce the inflammation marker CRP at the site of implantation." *Journal of Biomedical Materials Research* 61 (1). John Wiley & Sons, Ltd: 165–67. doi: 10.1002/jbm.10155.

Long, J., H. Kim, D. Kim, J. B. Lee, and D.-H. Kim. 2017. "A biomaterial approach to cell reprogramming and differentiation." *Journal of Materials Chemistry B* 5 (13). The Royal Society of Chemistry: 2375–89. doi: 10.1039/C6TB03130G.

Lu, J., H. Yu, and C. Chen. 2018. "Biological properties of calcium phosphate biomaterials for bone repair: A review." *RSC Advances* 8 (4). The Royal Society of Chemistry: 2015–33. doi:10.1039/C7RA11278E.

Luklinska, Z. B., and H. Schluckwerder. 2003. "*In vivo* response to HA-polyhydroxybutyrate/ polyhydroxyvalerate composite." *Journal of Microscopy* 211 (2). John Wiley & Sons, Ltd (10.1111): 121–29. doi: 10.1046/j.1365–2818.2003.01204.x.

Malik, A., M. H. Baig, and B. Manavalan. 2019. "Protein-carbohydrate interactions." In, edited by S. Ranganathan, M. Gribskov, K. Nakai, and C. Schönbach, *Encyclopedia of Bioinformatics and Computational Biology Schönbach*, 666–77. Oxford: Academic Press. doi: 10.1016/B978-0-12-809633-8.20661-4.

Mandal, B. B., S. Kapoor, and S. C. Kundu. 2009. "Silk fibroin/polyacrylamide semi-interpenetrating network hydrogels for controlled drug release." *Biomaterials* 30 (14). Netherlands: 2826–36. doi:10.1016/j.biomaterials.2009.01.040.

Mangir N, Roman S, Chapple CR, MacNeil S. Complications related to use of mesh implants in surgical treatment of stress urinary incontinence and pelvic organ prolapse: infection or inflammation?. *World Journal of Urology*. 38 (1). Springer 73-80.

Mano, J. F., G. A. Silva, H. S. Azevedo, P. B. Malafaya, R. A. Sousa, S. S. Silva, L. F. Boesel, et al. 2007. "Natural origin biodegradable systems in tissue engineering and regenerative medicine: Present status and some moving trends." *Journal of the Royal Society, Interface* 4 (17). The Royal Society: 999–1030. doi:10.1098/rsif.2007.0220.

Mao, A. S., and D. J. Mooney. 2015. "Regenerative medicine: Current therapies and future directions." *Proceedings of the National Academy of Sciences of the United States of America* 112 (47). United States: 14452–59. doi: 10.1073/pnas.1508520112.

Martínez-Martínez, M., G. Rodríguez-Berna, I. Gonzalez-Alvarez, M. a. J. Hernández, A. Corma, M. Bermejo, V. Merino, and M. Gonzalez-Alvarez. 2018. "Ionic hydrogel based on chitosan cross-linked with 6-phosphogluconic trisodium salt as a drug delivery system." *Biomacromolecules* 19 (4). ACS Publications: 1294–1304.

Masood, F., T. Yasin, and A. Hameed. 2015. "Polyhydroxyalkanoates–what are the uses? Current challenges and perspectives." *Critical Reviews in Biotechnology* 35 (4). Taylor & Francis: 514–21.

Matsuda, A., H. Kobayashi, S. Itoh, K. Kataoka, and J. Tanaka. 2005. "Immobilization of laminin peptide in molecularly aligned chitosan by covalent bonding." *Biomaterials* 26 (15). Elsevier: 2273–79.

McGregor, D. K. 1998. *From Midwives to Medicine: The Birth of American Gynecology.* Rutgers University Press 1–258.

McKellar, S.y. 2018. *Artificial Hearts: The Allure and Ambivalence of a Controversial Medical Technology.* JHU Press 1–350.

Melo, M. A. S., M. D. Weir, V. F. Passos, M. Powers, and H. H. Xu. 2017. "Ph-activated nanoamorphous calcium phosphate-based cement to reduce dental enamel demineralization." *Artificial Cells, Nanomedicine and Biotechnology* 45 (8). Informa UK Limited, trading as Taylor 8 Francis Group: 1778–85. doi:10.1080/21691401.2017.1290644.

Migonney, V. 2014. "History of Biomaterials." *Biomaterials*, 1–10. doi: 10.1002/9781119043553. ch1.

Miguel, S. P, M. P Ribeiro, H. Brancal, P. Coutinho, and I. J Correia. 2014. "Thermoresponsive chitosan–agarose hydrogel for skin regeneration." *Carbohydrate Polymers* 111. Elsevier: 366–73.

Millner, R. W. J., A. S. Lockhart, H. Bird, and C. Alexiou. 2009. "A new hemostatic agent: Initial life-saving experience with celox (chitosan) in cardiothoracic surgery." *The Annals of Thoracic Surgery* 87 (2). Elsevier: e13–14.

Mishra, M. 2018. *Encyclopedia of Polymer Applications, 3 Volume Set.* Boca Raton, FL. CRC Press.

Momoh, A. O, A. J. McKnight, A. Echo, S. E. Sharabi, J. C. Koshy, and L. H. Hollier Jr. 2010. "The first silicone breast implant patient: A 47-year follow-up." *Plastic and Reconstructive Surgery* 125 (6). LWW: 226e–229e.

Motealleh, A., S. Eqtesadi, A. Pajares, P. Miranda, D. Salamon, and K. Castkova. 2017. "Case study: Reinforcement of 45S5 bioglass robocast scaffolds by HA/PCL nanocomposite coatings." *Journal of the Mechanical Behavior of Biomedical Materials* 75 (July): 114–18. doi:10.1016/j.jmbbm.2017.07.012.

Bio-Nanomaterials

Mosahebi, A., P. Fuller, M. Wiberg, and G. Terenghi. 2002. "Effect of allogeneic schwann cell transplantation on peripheral nerve regeneration." *Experimental Neurology* 173 (2). Elsevier: 213–23.

Muddana, H. S., T. T. Morgan, J. H. Adair, and P. J. Butler. 2009. "Photophysics of Cy3-encapsulated calcium phosphate nanoparticles." *Nano Letters* 9 (4). American Chemical Society: 1559–66. doi:10.1021/nl803658w.

Murphy, A. R., and D. L. Kaplan. 2009. "Biomedical applications of chemically-modified silk fibroin." *Journal of Materials Chemistry* 19 (36): 6443–50. doi:10.1039/b905802h.

Murphy, S. V., and A. Atala. 2014. "3D Bioprinting of Tissues and Organs." *Nature Biotechnology* 32 (August). Nature Publishing Group, a division of Macmillan Publishers Limited. All Rights Reserved.: 773. doi: 10.1038/nbt.2958.

Muzzarelli, R. A. A., C. Zucchini, P. Ilari, A. Pugnaloni, M. Mattioli Belmonte, G. Biagini, and C. Castaldini. 1993. "Osteoconductive properties of methylpyrrolidinone chitosan in an animal model." *Biomaterials* 14 (12). Elsevier: 925–29.

Naskar, D., R. R. Barua, A. K. Ghosh, and S. C. Kundu. 2014. "Introduction to silk biomaterials." In *Silk Biomaterials for Tissue Engineering and Regenerative Medicine*, 3–40. Elsevier.

Natale, L. C., Y. Alania, M. C. Rodrigues, A. Simões, D. N. de Souza, E. de Lima, V. E. Arana-Chavez, et al. 2017. "Synthesis and characterization of silver phosphate/calcium phosphate mixed particles capable of silver nanoparticle formation by photoreduction." *Materials Science and Engineering* C 76. Elsevier B.V.: 464–71. doi:10.1016/j.msec.2017.03.102.

Ng, W. L., W. Y. Yeong, and M. W. Naing. 2016. "Polyelectrolyte gelatin-chitosan hydrogel optimized for 3D bioprinting in skin tissue engineering." *International Journal of Bioprinting* 2 (1).

Nitta, S. K., and Keiji N. 2013. "Biopolymer-based nanoparticles for drug/gene delivery and tissue engineering." *International Journal of Molecular Sciences* 14 (1): 1629–54. doi: 10.3390/ijms14011629.

No HK, Meyers SP. Preparation and characterization of chitin and chitosan—a review. Journal of aquatic food product technology. 4(2):taylor and francis: 27–52.

Novikov, L. N., L. N. Novikova, A. Mosahebi, M. Wiberg, G. Terenghi, and J.-O. Kellerth. 2002. "A novel biodegradable implant for neuronal rescue and regeneration after spinal cord injury." *Biomaterials* 23 (16). Elsevier: 3369–76.

Numata, K., and D. L. Kaplan. 2010. "Silk-based delivery systems of bioactive molecules." *Advanced Drug Delivery Reviews* 62 (15): 1497–1508. doi:10.1016/j.addr.2010.03.009.

Ooi, S. Y., I. Ahmad, and M. C. I. M. Amin. 2016. "Cellulose nanocrystals extracted from rice husks as a reinforcing material in gelatin hydrogels for use in controlled drug delivery systems." *Industrial Crops and Products* 93. Elsevier: 227–34.

Owens, G. J, R. K. Singh, F. Foroutan, M. Alqaysi, C.-M. Han, C. Mahapatra, H.-W. Kim, and J. C. Knowles. 2016. "Sol–gel based materials for biomedical applications." *Progress in Materials Science* 77: 1–79. doi: 10.1016/j.pmatsci.2015.12.001.

Padamwar, M. N., and A. P. Pawar. 2004. "Silk Sericin and Its Applications: A Review." CSIR.

Parhi, R. 2017. "Cross-linked hydrogel for pharmaceutical applications: A review." *Advanced Pharmaceutical Bulletin* 7 (4). Tabriz University of Medical Sciences: 515–30. doi: 10.15171/apb.2017.064.

Paridah, M.t., A. Moradbak, A. Z. Mohamed, F. a. t. Owolabi, M. Asniza, and S. H. P. A. Khalid. 2016. "We are IntechOpen, the world ' s leading publisher of open access books built by scientists, for scientists TOP 1%." *Intech* i (tourism): 13. doi: 10.5772/57353.

Park JB, Bronzino JD. Biomaterials: principles and applications. crc press; 2002, 1-249.

Park, J. B., and J. D. Bronzino. 2002. *Biomaterials: Principles and Applications*. Boca Raton, FL: CRC Press.

Park, S. J., T. W. Kim, M. K. Kim, S. Y. Lee, and S.-C. Lim. 2012. "Advanced bacterial polyhydroxyalkanoates: Towards a versatile and sustainable platform for unnatural tailor-made polyesters." *Biotechnology Advances* 30 (6). Elsevier: 1196–1206.

Park, W. H., R. W. Lenz, and S. Goodwin. 1998. "Epoxidation of bacterial polyesters with unsaturated side chains. I. production and epoxidation of polyesters from 10-undecenoic acid." *Macromolecules* 31 (5). ACS Publications: 1480–86.

Patil, A., and M. S. Ferritto. 2013. "Polymers for personal care and cosmetics: Overview." In *Polymers for Personal Care and Cosmetics* 1148: 1–3. ACS Symposium Series. American Chemical Society. doi:doi:10.1021/bk-2013-1148.ch001.

Patil, P. S., N. Fountas-Davis, H. Huang, M. M. Evancho-Chapman, J. A. Fulton, L. P. Shriver, and N. D. Leipzig. 2016. "Fluorinated methacrylamide chitosan hydrogels enhance collagen synthesis in wound healing through increased oxygen availability." *Acta Biomaterialia* 36. Elsevier: 164–74.

Pina, S., R. F. Canadas, G. Jimenez, M. Peran, J. A. Marchal, R. L. Reis, and J. M. Oliveira. 2017. "Biofunctional ionic-doped calcium phosphates: Silk fibroin composites for bone tissue engineering scaffolding." *Cells, Tissues, Organs* 204 (3–4). Switzerland: 150–63. doi: 10.1159/000469703.

Preda, R. C., G. Leisk, F. Omenetto, and D. L. Kaplan. 2013. "Bioengineered silk proteins to control cell and tissue functions." *Methods in Molecular Biology (Clifton, N.J.)* 996. United States: 19–41. doi:10.1007/978-1-62703-354-1_2.

Putthanarat, S., R. K. Eby, R. R. Naik, S. B. Juhl, M. A. Walker, E. Peterman, S. Ristich, et al. 2004. *Nonlinear Optical Transmission of Silk/Green Fluorescent Protein (GFP) Films. Polymer.* Vol. 45. doi:10.1016/j.polymer.2004.10.014.

Rahaiee, S., M. Zare, and S. M. Jafari. 2019. "Nanostructures of silk fibroin for encapsulation of food ingredients." In *Biopolymer Nanostructures for Food Encapsulation Purposes.* Elsevier: 305–31.

Rahmati, M., C. P. Pennisi, E. Budd, A. Mobasheri, and M. Mozafari. 2018. "Biomaterials for regenerative medicine: Historical perspectives and current trends." In *Cell Biology and Translational Medicine,* Vol. 4, 1–19. Springer.

Rai, R., T. Keshavarz, J. A. Roether, A. R. Boccaccini, and I. Roy. 2011. "Medium chain length polyhydroxyalkanoates, promising new biomedical materials for the future." *Materials Science and Engineering: R: Reports* 72 (3). Elsevier: 29–47.

Rajabian, M. H., G. H. Ghorabi, B. Geramizadeh, S. Sameni, and M. Ayatollahi. 2017. "Evaluation of bone marrow derived mesenchymal stem cells for full-thickness wound healing in comparison to tissue engineered chitosan scaffold in rabbit." *Tissue and Cell* 49 (1). Elsevier: 112–21.

Rakhshaei, R., and H. Namazi. 2017. "A potential bioactive wound dressing based on carboxymethyl cellulose/ZnO impregnated MCM-41 nanocomposite hydrogel." *Materials Science and Engineering: C* 73. Elsevier: 456–64.

Ramalingam, M., T. S. Sampath Kumar, S. Ramakrishna, and W. O. Soboyejo. 2016. *Biomaterials: A Nano Approach.* Boca Raton, FL: CRC press.

Ratner, B. D., A. S. Hoffman, F. J. Schoen, and J. E. Lemons. 2004. *Biomaterials Science: An Introduction to Materials in Medicine.* Elsevier 1–1573.

Rey, C., C. Combes, C. Drouet, and M. J. Glimcher. 2009. "Bone mineral: Update on chemical composition and structure." *Osteoporosis International* 20 (6): 1013–21. doi:10.1007/s00198-009-0860-y.

Rising, A., H. Nimmervoll, S. Grip, A. Fernandez-Arias, E. Storckenfeldt, D. P. Knight, F. Vollrath, and W. Engstrom. 2005. "Spider silk proteins—mechanical property and gene sequence." *Zoological Science* 22 (3). Japan: 273–81. doi:10.2108/zsj.22.273.

Rnjak-Kovacina, J., T. M. DesRochers, K. A. Burke, and D. L. Kaplan. 2015. "The effect of sterilization on silk fibroin biomaterial properties." *Macromolecular Bioscience* 15 (6): 861–74. doi:10.1002/mabi.201500013.

Bio-Nanomaterials

Ryu, S.-C., B.-K. Lim, F. Sun, K. Koh, D.-W. Han, and J. Lee. 2009. *Regeneration of a Tooth Enamel Layer Using Hydroxyapatite Regeneration of a Micro-Scratched Tooth Enamel Layer by Nanoscale Hydroxyapatite Solution. Bulletin of the Korean Chemical Society.* Vol. 30. doi:10.5012/bkcs.2009.30.4.887.

Sahoo, N. K., S. C. Anand, J. R. Bhardwaj, V. P. Sachdeva, and B. L. Sapru. 1994. "Bone response to stainless steel and titanium bone plates: An experimental study on animals." *Medical Journal, Armed Forces India* 50 (1). Elsevier: 10–14. doi:10.1016/S0377-1237(17)31029-8.

Salehi, M., S. Farzamfar, F. Bastami, and R. Tajerian. 2016. "Fabrication and characterization of electrospun PLLA/collagen nanofibrous scaffold coated with chitosan to sustain release of aloe vera gel for skin tissue engineering." *Biomedical Engineering: Applications, Basis and Communications* 28 (05). World Scientific: 1650035.

Salernitano, E., and C. Migliaresi. 2003. "Composite materials for biomedical applications: A review." *Journal of Applied Biomaterials and Biomechanics* 1 (1). SAGE Publications Sage UK: London, England: 3–18.

Sarkar, S. D., B. L. Farrugia, T. R. Dargaville, and S. Dhara. 2013. "Chitosan-collagen scaffolds with nano/microfibrous architecture for skin tissue engineering." *Journal of Biomedical Materials Research– Part A* 101 (12): 3482–92. doi: 10.1002/jbm.a.34660.

Sasaki, M., N. Kato, H. Watanabe, and H. Yamada. 2000. *Silk Protein, Sericin, Suppresses Colon Carcinogenesis Induced by 1,2-Dimethylhydrazine in Mice. Oncology Reports.* Vol. 7. doi:10.3892/or.7.5.1049.

Scholz, C. 2010. "Perspectives to produce positively or negatively charged polyhydroxyalkanoic acids." *Applied Microbiology and Biotechnology* 88 (4). Springer: 829–37.

Shah, M., M. I. Naseer, M. H. Choi, M. O. Kim, and S. C. Yoon. 2010. "Amphiphilic PHA–MPEG copolymeric nanocontainers for drug delivery: Preparation, characterization and *in vitro* evaluation." *International Journal of Pharmaceutics* 400 (1–2). Elsevier: 165–75.

Sheu, H.-S., K. W. Phyu, Y.-C. Jean, Y.-P. Chiang, I.-M. Tso, H.-C. Wu, J.-C. Yang, and S.-L. Ferng. 2004. "Lattice deformation and thermal stability of crystals in spider silk." *International Journal of Biological Macromolecules* 34 (5). Netherlands: 325–31. doi:10.1016/j.ijbiomac.2004.09.004.

Shim, K. S., S. E. Kim, Y. P. Yun, D. I. Jeon, H. J. Kim, K. Park, and H. R. Song. 2017. "Surface immobilization of biphasic calcium phosphate nanoparticles on 3D printed poly(caprolactone) scaffolds enhances osteogenesis and bone tissue regeneration." *Journal of Industrial and Engineering Chemistry* 55. The Korean Society of Industrial and Engineering Chemistry: 101–9. doi:10.1016/j.jiec.2017.06.033.

Sirichaisit, J., V. L. Brookes, R. J. Young, and F. Vollrath. 2003. "Analyis of structure/property relationships in silkworm (Bombyx mori) and spider dragline (Nephila edulis) silks using raman spectroscopy." *Biomacromolecules* 4 (2). United States: 387–94. doi:10.1021/bm0256956.

Sivashankari, P. R., and M. Prabaharan. 2016. "Prospects of chitosan-based scaffolds for growth factor release in tissue engineering." *International Journal of Biological Macromolecules* 93. Elsevier: 1382–89.

Smeriglio, P., J. H. Lai, F. Yang, and N. Bhutani. 2015. "3D hydrogel scaffolds for articular chondrocyte culture and cartilage generation." *JoVE (Journal of Visualized Experiments)* (104): 1–6.

Sodian, R., S. P. Hoerstrup, J. S. Sperling, D. P. Martin, S. Daebritz, J. E. Mayer Jr, and J. P. Vacanti. 2000. "Evaluation of biodegradable, three-dimensional matrices for tissue engineering of heart valves." *Asaio Journal* 46 (1). LWW: 107–10.

Stábile, F. M., M. P. Albano, L. B. Garrido, C. Volzone, P. T. De Oliveira, and A. L. Rosa. 2016. "Processing of ZrO2 scaffolds coated by glass–ceramic derived from 45S5 bioglass." *Ceramics International* 42 (3): 4507–16. doi: 10.1016/j.ceramint.2015.11.140.

Stupp, S. I., J. J. J. M. Donners, L.-s. Li, and A. Mata. 2005. "Expanding Frontiers in Biomaterials." *Mrs Bulletin* 30 (11). Cambridge University Press: 864–73.

Sudheesh Kumar, P. T., V.-K. Lakshmanan, T. V. Anilkumar, C. Ramya, P. Reshmi, A. G. Unnikrishnan, S. V. Nair, and R. Jayakumar. 2012. "Flexible and microporous chitosan hydrogel/nano ZnO composite bandages for wound dressing: *In vitro* and *in vivo* evaluation." *ACS Applied Materials & Interfaces* 4 (5). ACS Publications: 2618–29.

Sun, J., D. Wei, K. Yang, Y. Yang, X. Liu, H. Fan, and X. Zhang. 2017. "The development of cell-initiated degradable hydrogel based on methacrylated alginate applicable to multiple microfabrication technologies." *Journal of Materials Chemistry B* 5 (40). The Royal Society of Chemistry: 8060–69. doi: 10.1039/C7TB01458A.

Suzuki, K., T. Mikami, Y. Okawa, A. Tokoro, S. Suzuki, and M. Suzuki. 1986. "Antitumor effect of hexa-N-acetylchitohexaose and chitohexaose." *Carbohydrate Research* 151. Elsevier: 403–8.

Tamada, Y., M. Sano, K. Niwa, T. Imai, and G. Yoshino. 2004. "Sulfation of silk sericin and anticoagulant activity of sulfated sericin." *Journal of Biomaterials Science. Polymer Edition* 15 (8). England: 971–80.

Tang, F., L. Lv, F. Lu, B. Rong, Z. Li, B. Lu, K. Yu, J. Liu, F. Dai, and D. Wu. 2016. "Preparation and characterization of N-chitosan as a wound healing accelerator." *International Journal of Biological Macromolecules* 93. Elsevier: 1295–1303.

Tara, S., K.A. Rocco, N. Hibino, T. Sugiura, H. Kurobe, C. K. Breuer, and T. Shinoka. 2014. "Vessel bioengineering." *Circulation Journal : Official Journal of the Japanese Circulation Society* 78 (1). Japan: 12–19.

Teo, A. J. T., A. Mishra, I. Park, Y.-J. Kim, W.-T. Park, and Y.-J. Yoon. 2016. "Polymeric biomaterials for medical implants and devices." *ACS Biomaterials Science & Engineering* 2 (4). ACS Publications: 454–72.

Thi, P. L., Y. Lee, T. T. H. Thi, K. M. Park, and K. D. Park. 2018. "Catechol-rich gelatin hydrogels in situ hybridizations with silver nanoparticle for enhanced antibacterial activity." *Materials Science and Engineering: C* 92. Elsevier: 52–60.

Thurber, A. E., F. G. Omenetto, and D. L. Kaplan. 2015. "In vivo bioresponses to silk proteins." *Biomaterials* 71 (December): 145–57. doi:10.1016/j.biomaterials.2015.08.039.

Tibbitt, M. W, and K. S. Anseth. 2009. "Hydrogels as extracellular matrix mimics for 3D cell culture." *Biotechnology and Bioengineering* 103 (4). United States: 655–63. doi: 10.1002/bit.22361.

Tokoro, A., N. Takewaki, K. O. Suzuki, T. Mikami, S. Suzuki, and M. Suzuki. 1988. "Growth-inhibitory effect of hexa-N-acetylchitohexanse and chitohexaose against meth-A solid tumor." *Chemical and Pharmaceutical Bulletin* 36 (2). The Pharmaceutical Society of Japan: 784–90.

Tozzi, P. 2007. *Sutureless Anastomoses: Secrets for Success*. Springer 1–137.

Trivedi, R. H., D. J. Apple, S. K. Pandey, L. Werner, A. M. Izak, A. R. Vasavada, and J. Ram. 2003. "Sir Nicholas Harold Ridley. He changed the world, so that we might better see it." *Indian Journal of Ophthalmology* 51 (3). Medknow Publications: 211.

Ueno, H., T. Mori, and T. Fujinaga. 2001. "Topical formulations and wound healing applications of chitosan." *Advanced Drug Delivery Reviews* 52 (2). Elsevier: 105–15.

Villanueva, M. Emilia, A. M. d. R. Diez, J. A. González, C. J. Pérez, M. Orrego, L. Piehl, S. Teves, and G. J. Copello. 2016. "Antimicrobial activity of starch hydrogel incorporated with copper nanoparticles." *ACS Applied Materials & Interfaces* 8 (25). ACS Publications: 16280–88.

Wang, J., and J. Wei. 2017. "Interpenetrating network hydrogels with high strength and transparency for potential use as external dressings." *Materials Science and Engineering: C* 80. Elsevier: 460–67.

Wang, L., E. Khor, and L.-Y. Lim. 2001. "Chitosan–alginate–$CaCl_2$ system for membrane coat application." *Journal of Pharmaceutical Sciences* 90 (8). Wiley Online Library: 1134–42.

Wang, L., E. Khor, A. Wee, and L. Y. Lim. 2002. "Chitosan-alginate PEC membrane as a wound dressing: Assessment of incisional wound healing." *Journal of Biomedical Materials Research: An Official Journal of The Society for Biomaterials, The Japanese Society for Biomaterials, and The Australian Society for Biomaterials and the Korean Society for Biomaterials* 63 (5). Wiley Online Library: 610–18.

Wang, Y.-W., Q. Wu, and G.-Q. Chen. 2004. "Attachment, proliferation and differentiation of osteoblasts on random biopolyester poly (3-hydroxybutyrate-Co-3-hydroxyhexanoate) scaffolds." *Biomaterials* 25 (4). Elsevier: 669–75.

Wang, Z., Y. Itoh, Y. Hosaka, I. Kobayashi, Y. Nakano, I. Maeda, F. Umeda, J. Yamakawa, M. Nishimine, and T. Suenobu. 2003. "Mechanism of enhancement effect of dendrimer on transdermal drug permeation through polyhydroxyalkanoate matrix." *Journal of Bioscience and Bioengineering* 96 (6). Elsevier: 537–40.

Webb, J. C. J., and R. F. Spencer. 2007. "The role of polymethylmethacrylate bone cement in modern orthopaedic surgery." *The Journal of Bone and Joint Surgery. British Volume* 89 (7). The British Editorial Society of Bone and Joint Surgery: 851–57.

Weir, M. D., J. Ruan, N. Zhang, L. C. Chow, K. Z., X. Chang, Y. Bai, and H. H. Xu. 2017. "Effect of calcium phosphate nanocomposite on in vitro remineralization of human dentin lesions." *Dental Materials* 33 (9). The Academy of Dental Materials: 1033–44. doi:10.1016/j.dental.2017.06.015.

Wong, J. Y., D. R. Peterson, J. D. Bronzino. 2012. Biomaterials: Principles and Practices. https://books.google.com/books?id=mE9iK9UibNAC&pgis=1.

Woodard, J. R, A. J. Hilldore, S. K. Lan, C. J. Park, A. W. Morgan, J. A. C. Eurell, S. G. Clark, M. B. Wheeler, R. D. Jamison, and A. J. W. Johnson. 2007. "The mechanical properties and osteoconductivity of hydroxyapatite bone scaffolds with multi-scale porosity." *Biomaterials* 28 (1). Elsevier: 45–54.

Wu, Q., Y. Wang, and G.-Q. Chen. 2009. "Medical application of microbial biopolyesters polyhydroxyalkanoates." *Artificial Cells, Blood Substitutes, and Biotechnology* 37 (1). Taylor & Francis: 1–12.

Xia, Y., H. Chen, F. Zhang, L. Wang, B. Chen, M. A. Reynolds, J. Ma, A. Schneider, N. Gu, and H. H. Xu. 2018. "Injectable calcium phosphate scaffold with iron oxide nanoparticles to enhance osteogenesis via dental pulp stem cells." *Artificial Cells, Nanomedicine and Biotechnology* 0 (0). Informa UK Limited, trading as Taylor & Francis Group: 1–11. doi:10.1080/21691401.2018.1428813.

Xin, Y., M. Yin, L. Zhao, F. Meng, and L. Luo. 2017. "Recent progress on nanoparticle-based drug delivery systems for cancer therapy." *Cancer Biology & Medicine* 14 (3). Chinese Anti-Cancer Association: 228–41. doi: 10.20892/j.issn.2095-3941.2017.0052.

Xu, H. H., J. L. Moreau, L. Sun, and L. C. Chow. 2011. "Nanocomposite containing amorphous calcium phosphate nanoparticles for caries inhibition." *Dental Materials : Official Publication of the Academy of Dental Materials* 27 (8). England: 762–69. doi:10.1016/j.dental.2011.03.016.

Xu, S., X. Yang, X. Chen, H. Shao, Y. He, L. Zhang, G. Yang, and Z. Gou. 2014. "Effect of borosilicate glass on the mechanical and biodegradation properties of 45S5-derived bioactive glass-ceramics." *Journal of Non-Crystalline Solids* 405: 91–99. doi: 10.1016/j. jnoncrysol.2014.09.002.

Yadav, T. C., A. K. Srivastava, N. Raghuwanshi, N. Kumar, R. Prasad, and V. Pruthi. 2019. "Wound healing potential of natural polymer: Chitosan 'a wonder molecule.'" *Integrating Green Chemistry and Sustainable Engineering*. Hoboken, NJ: John Wiley & Sons, Inc, 527–79.

Yan, X., E. Khor, and L.-Y. Lim. 2000. "PEC films prepared from chitosan-alginate coacervates." *Chemical and Pharmaceutical Bulletin* 48 (7). The Pharmaceutical Society of Japan: 941–46.

Yang, W., E. Fortunati, F. Bertoglio, J. S. Owczarek, G. Bruni, M. Kozanecki, J. M. Kenny, L. Torre, L. Visai, and D. Puglia. 2018. "Polyvinyl alcohol/chitosan hydrogels with enhanced antioxidant and antibacterial properties induced by lignin nanoparticles." *Carbohydrate Polymers* 181. Elsevier: 275–84.

Ye, C., P. Hu, M.-X. Ma, Y. Xiang, R.-G. Liu, and X.-W. Shang. 2009. "PHB/PHBHHx scaffolds and human adipose-derived stem cells for cartilage tissue engineering." *Biomaterials* 30 (26). Elsevier: 4401–6.

Yih, T. C., and M. Al-Fandi. 2006. "Engineered nanoparticles as precise drug delivery systems." *Journal of Cellular Biochemistry* 97 (6). United States: 1184–90. doi: 10.1002/jcb.20796.

Yoon, D. S., Y. Lee, H. A. Ryu, Y. Jang, K.-M. Lee, Y. Choi, W. J. Choi, M. Lee, K. M. Park, and K. D. Park. 2016. "Cell recruiting chemokine-loaded sprayable gelatin hydrogel dressings for diabetic wound healing." *Acta Biomaterialia* 38. Elsevier: 59–68.

Zhang, D., W. Zhou, B. Wei, X. Wang, R. Tang, J. Nie, and J. Wang. 2015. "Carboxyl-modified poly (vinyl alcohol)-crosslinked chitosan hydrogel films for potential wound dressing." *Carbohydrate Polymers* 125. Elsevier: 189–99.

Zhang, H., X. Luo, H. Tang, M. Zheng, and F. Huang. 2017a. "A novel candidate for wound dressing: Transparent porous maghemite/cellulose nanocomposite membranes with controlled release of doxorubicin from a simple approach." *Materials Science and Engineering: C* 79. Elsevier: 84–92.

Zhang, X., J. Li, P. Ye, G. Gao, K. Hubbell, and X. Cui. 2017b. "Coculture of mesenchymal stem cells and endothelial cells enhances host tissue integration and epidermis maturation through AKT activation in gelatin methacryloyl hydrogel-based skin model." *Acta Biomaterialia* 59. Elsevier: 317–26.

Zhang, Y.-Q., Y. Ma, Y.-Y. Xia, W.-D. Shen, J.-P. Mao, and R.-Y. Xue. 2006. "Silk sericin-insulin bioconjugates: Synthesis, characterization and biological activity." *Journal of Controlled Release : Official Journal of the Controlled Release Society* 115 (3). Netherlands: 307–15. doi:10.1016/j.jconrel.2006.08.019.

Zhao, K., Y. Deng, J. C. Chen, and G.-Q. Chen. 2003. "Polyhydroxyalkanoate (PHA) scaffolds with good mechanical properties and biocompatibility." *Biomaterials* 24 (6). Elsevier: 1041–45.

Zhao, Z., Y. Li, and M.-B. Xie. 2015. "Silk fibroin-based nanoparticles for drug delivery." *International Journal of Molecular Sciences* 16 (3). MDPI: 4880–903. doi:10.3390/ijms16034880.

Zhu, Y., K. Zhang, R. Zhao, X. Ye, X. Chen, Z. Xiao, X. Yang, et al. 2017. "Bone regeneration with micro/nano hybrid-structured biphasic calcium phosphate bioceramics at segmental bone defect and the induced immunoregulation of MSCs." *Biomaterials* 147. Elsevier Ltd: 133–44. doi:10.1016/j.biomaterials.2017.09.018.

Zinn, M., B. Witholt, and T. Egli. 2001. "Occurrence, synthesis and medical application of bacterial polyhydroxyalkanoate." *Advanced Drug Delivery Reviews* 53 (1). Elsevier: 5–21.

9 Biomaterials

Manoj Mittal
IKG Punjab Technical University

CONTENTS

9.1 History of Biomaterials 166
9.2 Orthopaedic and Dental Implant Materials................................... 166
    9.2.1 Metallic Implant Materials... 166
    9.2.2 Biomaterials Based on Ca and P... 169
        9.2.2.1 Tri-Calcium Phosphate (TCP) 171
        9.2.2.2 Tetracalcium Phosphate (TTCP)............................ 171
        9.2.2.3 Amorphous Calcium Phosphate (ACP) 172
        9.2.2.4 Apatite.. 172
        9.2.2.5 Porous and Dense HA Materials 173
        9.2.2.6 HA-Based Composites... 174
        9.2.2.7 HA-Based Composite Coatings 174
        9.2.2.8 Bond Coat 177
        9.2.2.9 Significance of Bone Implant Interface 178
        9.2.2.10 Comparison of Biological and Synthetic HA 181
9.3 Failure Mechanism of HA Coatings... 181
    9.3.1 Dissolution Behaviour of Hydroxyapatite........................... 182
    9.3.2 Unresolved Issues of Degradation of Hydroxyapatite 186
References.. 187

Any material that concoct to co-operate with the living organism for a medicinal determination, either a healing (treat, augment, repair or replace a tissue function of the body) or a diagnostic, is known as biomaterial. Due to the increased expected life of human being, various new materials are discovered and investigated with better service life in harsh body environment. From inserting intraocular lens in eye for removal of cataract to repairing heart blockage to replace broken bone, everywhere biomaterials are put into service. These can be classified depending upon their use, their service life, tendency for cell ingrowth, load-bearing capability, tendency of dissolution and many more. Gold, silver and alumina are the biomaterials used for repair and replacement of teeth since long. Metal and their alloys, high density polymers and derivatives of glass and calcium and phosphorus (Ca-P) are different categories of biomaterials. Various types of biomaterials are discussed in this chapter with respect to their utilization in different parts of human body and what is its expected service life.

165

9.1 HISTORY OF BIOMATERIALS

Use of materials for living organism has been recorded in pre-history. The remains of human found near Kennewick, Washington, United States, were supposed to be as old as 9000 years. As designated by archaeologist he was giant, fit, energetic person that walked through southern Washington, with a javelin point embedded in his hip. Spearpoint heeled in his body and did not obstruct his activities. This accidental insert explains the capacity of body to deal with extraneous things. The insert has a little similarity with today's implants, but it was a tolerated extraneous thing.

Mayan person often created nacre teeth from seashells roughly 600 AD and apparently accomplished what we now state as integration of bone, it is nothing but a seamless addition to the tissues. Similarly, a dental implant made up of iron dated 200 AD was found in Europe. No material science, biological understanding or medicine used for these procedures, still, their success highlights two points: the lenient nature of surrounding tissues and the pressing drive.

Evidence of use of sutures (seams or joints) is about 32,000 years old. Large wounds were closed early history by cautery (technique of burning a part of a body to remove or close off a part of it) or sutures. Latin sutures were commonly used by early Egyptians. Catgut was used in Europe, while metallic sutures in Greek. Galen from Pergamon called ligature made of precious metal. Use of led wire sutures with little reaction is also reported. A lot of problems have been faced with sutures in eras, where sterilization, toxicology, immunological, reaction to extraneous biological materials, inflammation and biodegradation are itself a big issue. Yet, sutures were relatively common fabricated or manufactured biomaterial for thousands of years.

9.2 ORTHOPAEDIC AND DENTAL IMPLANT MATERIALS

Biomaterials refer to synthetic materials used to replace a part of a living system or to function in intimate contact with living tissue. From the nature of materials point of view: biomaterials can be classified as polymers, ceramics, metals and composites. This classification may be overlaid on each other due to the fact that composites are made from two or more types (Park and Lakes, 2007). So far, various kinds of biomaterials have been used in humans as reported in Table 9.1.

9.2.1 METALLIC IMPLANT MATERIALS

The majority of metallic implants used in the skeletal system are based on three types of alloy systems: austenitic stainless steels, cobalt–chromium alloys and titanium-based alloys (Bonfield, 1987;). The fixation of these metallic implants to bone has been one of the most difficult problems, and the fixation is obtained through the acrylic bone cement. Cobalt–chromium (Co-Cr) alloys show excellent wear resistance properties, corrosion resistance and fatigue resistance, although they have higher elastic moduli and are less biocompatible than titanium as reported in Table 9.2. A process where the stiff implant carries much higher load than the surrounding bone called stress shielding is found to be increased due to high modulus of elasticity of Co-Cr alloys. In addition, over 10% of the population that goes for body implant is

Biomaterials

TABLE 9.1
Biomaterials Used in Body System (Zhang, 2000)

System	Components
Skeletal	Bone plate, total joint replacement
Muscular	Sutures
Digestive	Sutures
Circulatory	Artificial heart valves, blood vessels
Respiratory	Oxygenator machine
Integumentary	Sutures, burn dressings, artificial skin
Urinary	Catheters, kidney dialysis machine
Nervous	Hydrocephalus drain, cardiac pacemaker
Endocrine	Microencapsulated pancreatic islet cells
Reproductive	Augmentation mammoplasty, other cosmetic replacements

TABLE 9.2
Properties of Biomaterials Used in Implants (Long and Rack, 1998)

Materials	Elastic Modulus $(Pa \times 10^9)$	Tensile Strength $(Pa \times 10^6)$	Advantages	Disadvantages
cp Ti	110	400	Biocompatible, corrosion resistance	Fatigue resistance, wear resistance
Ti-6Al-4V	124	940	Biocompatible, corrosion resistance, fatigue strength	Wear resistance (fretting)
SS 316L	193	540	Low cost, easy availability and processing	Poor biocompatibility, corrosion, high modulus
Co-Cr	214	480	Wear and resistance to corrosion, fatigue strength	High modulus, meager biocompatibility
UHMWPE	~0.4	2–4	Low friction and low wear	Susceptible to 3rd body wear and particle creation
PMMA	2–4	30–55	Rapid fixation, strong fixation for short time	Low impact and fatigue resistance, poor biocompatibility
Bone	12–25	75–140	---	---

cpTi, commercially pure titanium; UHMWPE, ultra-high molecular weight polyethylene; PMMA, polymethylmethacrylate.

hypersensitive to at least one of the elements in Co-Cr alloys (9.6% Ni, 9.3% Cr and 6.0% Co), Ti alloy (4% V) and stainless steels (10% Ni and 16% Cr) (Black, 1999; Browne and Gregson, 1995).

The current implant designs are one of the pioneer works conducted by Sir John Charnley (Todd et al., 1972). A Co-Cr-Mo alloy (ASTM F75), Co-Ni-Cr-Mo alloy (ASTM F562) or Ti-Al-V alloy (ASTM F136) surface articulates with the surface of

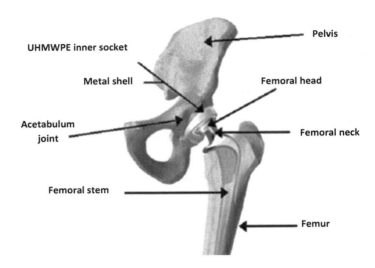

FIGURE 9.1 Modelled total hip prosthesis implant (Johnson, 2005).

ultra-high molecular weight polyethylene (UHMWPE) in a total joint replacement as presented in Figure 9.1. Polymethylmethacrylate (PMMA) bone cement is used to fix these components in place (Zhang, 2000).

The combination of metallic femoral head and UHMWPE inner socket used in current implant devices has exceedingly short life span. It is believed that about 20% of implants must be substituted within 10 years, and 3%–6% may be replaced within 5 years of installation (Birtwistle et al., 1996). Slackening, tribology, decomposition, irregular stress distributions and tissue inflammation contribute to short service life of metallic body implants. The comparatively poor durability of prostheses due to inadequate adhesive expertise encouraged by Charnley to be used in old age patients with a low life hope (Emery et al., 1997). Early problems with implants, such as infection and stem breakage, are largely avoided today, but the long-term failure of a graft is most often credited to implant slackening, which can cause severe pain, loss of joint function that necessitate revision surgery (Netter et al., 1989, Spector, 1990, Spector et al., 1991; Long and Rack, 1998).

Hench (1998) reported that the failure analysis has made people reconsider the use of acrylic bone cement at bone-implant junction since it is progressive deterioration at the interface that results in failure of either bone or implant. Consequently, there is a growing interest in alternative, cement-less methods for fastening implants with bone. Three primary methods of cement-less procedures have been developed: morphological, biological (porous in-growth of bone) and bioactive fixation (surface-active interfacial bonding). Morphological fixation is intended to anchor the implant in bone by mechanical interlocking and spread the load by increasing the surface area of the implant. Absence of cement damage to bone, if revision surgery is necessary, is a potential clinical advantage for morphological fixation of implant devices. However, revision surgeries are complicated due to bone and tissue in-growth into indentations and superficial irregularities. The in-growth of bone into porous insert

Biomaterials 169

is the principle for biological fixation. The load can be transmitted from implant to bone over a large interfacial area, and localized stress gradients can be reduced. Unlike the cement-based fixation of implant interface, such a living bone-implant interface would remodel in response to stress variation over very long period and interfacial loosening should be delayed or prevented. However, it has been found that interfacial strength decreases during long period of time.

Ideally, implants should be biocompatible as well as bioactive, which means that the implants can be osteoconductive or osteoinductive, i.e. they promote new bone growth around the implants (Mittal et al. 2013). Bioactive fixation requires implant materials to be bioactive, which can establish bone bonding with living tissue by physiochemical or cell-mediated biological process. There are two types of bioactive materials currently in use for orthopaedic and dental applications: bio-glass and calcium phosphates. Bio-glass was invented by Hench in the late 1960s. He first developed a surface reactive bio-glass, containing SiO_2, Na_2O, CaO and P_2O_5, which can be chemically bonded to bone. Nowadays, this type of bio-glass is widely used in clinical applications. Further, with very good bioactive properties of these materials, the mechanical properties are very poor.

In a study of radiographic examination of hip arthroplasty patients after 10 years, 30%–40% of the patients showed femoral loosening and 10%–30% showed acetabular loosening. After 15 years of implantation, the rates of failure were much higher (Netter et al., 1989). Implant loosening results from mechanical failure of supporting cancellous bone due to bone resorption (osteolysis) and break down of material tissue interface, but the cascade of events leading to it is complex and can take many years to develop (Spector, 1990, and Spector et al., 1991). Flowchart of many actions that may cause failure of implant is reported in Figure 9.2.

At the root of these problems, there are two major causes: osteolytic response to wear particles generated from the articulating surface and osteolytic response due to poor long-term biocompatibility and/or loss of fixation at bone–biomaterial interface. Many of these problems depend fundamentally on material selection and design of implant.

9.2.2 Biomaterials Based on Ca and P

Calcium phosphates (Ca-P) are integral constituent of geographical and living systems. Ca-P is found in rocks and is the major foundations of animals. Calcium and phosphate are extensively dispersed elements on earth. The top-most layer of the crest of this planet contains 3.4% Ca and 0.10% P, respectively (Weast, 1985–1986). Table 9.3 presents most of the compounds made from calcium and phosphorous ceramics and their Ca/P ratio. The most widely used Ca-P ceramics for biomedical purpose are hydroxyapatite (HA), tri-calcium phosphate and their biphasic combinations (Le Geros, 1991, Chow, 2000, Suchanek and Yoshimura, 1998; Vallet-Regi and Gozalez-Calbet, 2004). Combinations of oxides of calcium and phosphorus with or without water content give different calcium phosphates. All calcium phosphates are only sparingly dissolvable in water, and some may be considered non-dissolvable but all dissolves in acids.

Calcium orthophosphates are of interest as they are the main components of human fossilized tissues. Hence Ca-P is of great importance in the field of natural

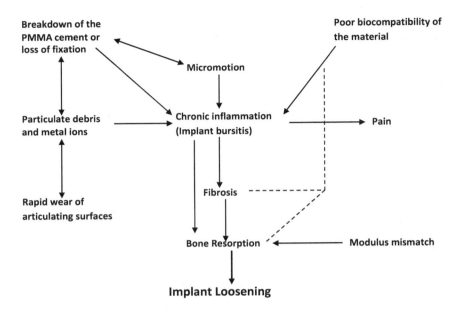

FIGURE 9.2 Flowchart showing the material-related causes and the intermediate events leading up to implant loosening (Bell, 2004).

TABLE 9.3
Abbreviations, Formula and Ca/P Ratio of Calcium Phosphorus Ceramics (Liven, 2007, Shi, 2004; Levingstone, 2008)

Compound Name	Abbreviation	Formula	Ca/P Ratio
Mono-calcium phosphate monohydrate	MCPM	$Ca(H_2PO_4)_2 \cdot H_2O$	0.5
Mono-calcium phosphate anhydrous	MCPA	$Ca(H_2PO_4)_2$	0.5
Dicalcium phosphate dehydrate	DCPD	$CaHPO_4 \cdot 2H_2O$	1.0
Dicalcium phosphate anhydrate	DCPA	$CaHPO_4$	1.0
Octacalcium phosphate	OCP	$Ca_8H_2(PO_4)_6 \cdot 5H_2O$	1.33
α-tricalcium phosphate	α-TCP	$\alpha\text{-}Ca_3(PO_4)_2$	1.5
β-tricalcium phosphate	β-TCP	$\beta\text{-}Ca_3(PO_4)_2$	1.5
Amorphous calcium phosphate	ACP	$Ca_x(PO_4)_y \cdot nH_2O$	1.2 – 1.5
Calcium deficient hydroxyapatite	CDHA	$Ca_{10-x}(HPO_4)_x(PO_4)_{6-x}(OH)_{2-x}$	1.5 – 1.67
Hydroxyapatite	HA	$Ca_{10}(PO_4)_6(OH)_2$	1.67
Tetracalcium phosphate	TTCP	$Ca_4(PO_4)_2°$	2.0

science, medicine, dentistry, geology, etc. Their structure, composition, solubility and stability are responsible for their formation, function and applications. In vivo formation of inorganic minerals is known as biomineralization. More than 60 different biominerals have been identified for use in medicine industry, main constituents of which are H, C, O, Mg, P, S, Ca, Mn and Fe elements (Shi, 2004).

Biomaterials 171

Numerous types of minerals of phosphate co-exist in some pathological tissue calcification, e.g. dicalcium phosphate dihydrate (DCPD), octacalcium phosphate (OCP) (Holt, 2009), tricalcium phosphate (TCP) and apatite in dental calculi, urinary and salivary stones (Mann, 2001). In some cases of pathological tissue calcification, the phosphate minerals exist with non-phosphate minerals. The general occurrence and existence of phosphate minerals in human tissues are summarized in Table 9.4.

9.2.2.1 Tri-Calcium Phosphate (TCP)

β-Tri-calcium phosphate (β-TCP) is one of the calcium phosphates with stoichiometric composition $Ca_3(PO_4)_2$, having rhombohedral structure that cannot be precipitated from solution but can be prepared by the calcification of calcium-deficient HA at a temperature more than 800°C by following chemical reaction:

$$Ca_9(HPO_4)(PO_4)_5 OH \rightarrow \underset{(TCP)}{3Ca_3(PO_4)_2} + H_2O \qquad (9.1)$$

At 1125°C, β-TCP transforms into α-TCP (high melting temperature) phase, β-TCP is less soluble in water than α-TCP. Pure β-TCP never occurs in organic calcification. Whitlockite (β-(Ca, Mg)$_3$(PO$_4$)$_2$), magnesium-containing form is found in dental calculi, urinary stone, dental caries salivary stone, arthritic cartilage and some soft tissue deposits (Le Geros, 1994). α-TCP is a metastable phase at room temperature (Dorozhkin and Epple, 2002).

9.2.2.2 Tetracalcium Phosphate (TTCP)

Tetracalcium phosphate (TTCP) is a monoclinic phase, which dissolves easily in water as compared to HA. TTCP can only be prepared by a solid-state reaction above 1300°C by following chemical reaction.

TABLE 9.4

Phosphate Minerals and Their Occurrence in Human Body (Le Geros and Le Geros, 1984)

Minerals	Chemical Formula	Occurrences
Apatite or apatitic calcium phosphates	(Ca, Na, Sr, K)$_{10}$ (PO$_4$, CO$_3$, HPO$_4$)$_6$ (OH, F, Cl)$_2$	Enamel, dentine, bone, salivary stone, dental calculi, soft-tissue calcification
Whitlockite	(Ca, Mg)$_9$(PO$_4$)$_6$	Salivary stone, dental calculi, calcified cartilage
Octacalcium phosphate	Ca$_8$H$_2$(PO$_4$)$_6$ ·5H$_2$O	Dental and urinary calculi
Brushite	CaHPO$_4$ ·2H$_2$O	Dental calculi, concretions in old bones, chondrocalcinosis
Calcium pyrophosphate dehydrate	Ca$_2$P$_2$O$_7$ ·2H$_2$O	Pseudo-gout deposits in synovium fluid
Struvite	Mg NH$_4$PO$_4$ ·6H$_2$O	Urinary stone
Newberyite	MgHPO$_4$ ·3H$_2$O	Urinary stone
Amorphous calcium phosphate	Variable composition	Non-visceral calcifications associated with uraemia

$$2CaHPO_4 + 2CaCO_3 \rightarrow \underset{\text{(TTCP)}}{Ca_4(PO_4)_2O} + 2CO_2 + H_2O \tag{9.2}$$

TTCP is not very stable in aqueous solution and slowly hydrolyses to HA and calcium hydroxide.

9.2.2.3 Amorphous Calcium Phosphate (ACP)

HA in its crystalline form is supposed to be a stable product in the precipitation of ions of Ca and P from the solutions. The morphology of amorphous calcium phosphate (ACP) consists of roughly spherical $Ca_9(PO_4)_6$ clusters combined arbitrarily with their inter-group spaces filled with water. Sedlak and Beebe (1974) reported that temperature programmed-dehydroxylation of ACP indicates that about 75% of this water is tightly bound inside the ACP particles. The rest (25%) is more loosely held surface water. The transition of ACP to crystalline occurs sharply, which is sensitive to temperature and pH (Boskey and Posner, 1973). The alteration process takes place by propagation of small crystals with a dendrite-like growing mechanism. The concentration level falls off sharply at pH < 9.25.

9.2.2.4 Apatite

The term apatite defines three unique minerals: hydroxy, fluoro and chlorapatite with OH ions replaced with F and Cl ions, respectively. The atomic arrangement of the three apatite phases differs due to position of the occupation of [00z] anion position i.e. hydroxyl, fluorine and chlorine groups, for three end members, respectively (Hughes and Rakovan, 2002). These materials demonstrate a similar structure and possess the structural formula $X_3Y_2(TO_4)Z$. This structure allows easy substitution of apatite, which includes X = Y = (Ca), Sr, Ba, Re, Pb, U or Mn (rarely Na, K, Y, Cu); T = (P), As, V, Si, S or C (as CO_3) and Z = (F), Cl, (OH) or O (medicine). Out of these three apatites, HA is the most studied and used biomaterial for orthopaedic and dental implants. HA has a crystalline structure of hexagonal rhombic prism as shown in Figure 9.3 with $a = 9.432$Å and $c = 6.881$Å (lattice parameters). Hydroxyl (OH^-) ions occur at the corners of the basal plane. These ions are positioned at every one half of the unit cell measured 3.44Å, parallel to c-axis and perpendicular to basal plane. Sixty percent of the calcium ions in the unit cell are associated with the hydroxyl ions (Johnson, 2005). The melting point (mp) of HA is 1550°C. The ideal Ca/P ratio of HA is 10/6 ≈ 1.67 and calculated density is 3.219×10^3 kg/m^3. HA is found in hard tissues, such as bone, dentine and dental enamel (Parks and Lakes, 1992). HA has excellent bioactivity and biocompatibility and can make natural union (chemical bonding) with hard tissues. HA is an osteoconductive material that provides temporary scaffold for bone growth. The osseointegration of a biomaterial depends on the properties of implant material, surface conditions, bone status, implantation technique and loading conditions on implant.

For clinical use, HA is available in various forms e.g. HA powders, porous HA, HA-based composites and HA-based coatings. Its Young's modulus (E), Poisson's ratio (ν), Vickers hardness (H_v) and bending strength (σ) were reported to be 40–117 GPa, 0.27 GPa, 3.43 GPa and 147 MPa, respectively.

Biomaterials 173

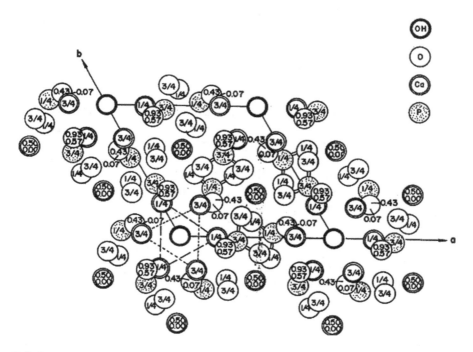

FIGURE 9.3 Structure of hydroxyapatite projected down the c-axis onto basal plane (Zhang, 2000).

9.2.2.5 Porous and Dense HA Materials

A porous HA ceramic enhances chemical bonding with bone, and interconnecting pore network improves the circulation of nutrients deep inside the bone. The porosity increases solubility at the cost of mechanical properties. Thus, more porous HA cannot be heavily loaded and are used only for defects. If size of pores is less than 5 μm, the material is called microporous, and if size of pores is more than 100 μm, the material is called macro-porous. Microporous material is important for bioresorbability while macro-porous for osteoconductive, respectively. Optimum macroporosity for in-growth of bone tissue is between 150 and 500 μm.

In some studies, strontium presented improvement in the integration of converted corals in bones. A complete transformation to apatite by hydrothermal treatment in presence of potassium dihydrogen phosphate has been reported (Xu et al., 2001).

Pure and dense material can be produced through sintering process, in which calcification and compaction are done at room temperature followed by high-temperature heating. While heating in temperature range of 600–900°C, moisture, carbonates and chemicals (which are present since initial stage) and ammonia and nitrates are removed as gaseous products, leading to production of dense HA material during sintering. The increased size of crystal and decreased surface area per unit volume are the result of processing. The product with a Ca/P ratio more than 1.67 and presence of CaO is the result of sintering. The presence of calcium oxide decreases the cohesion strength (due to stresses arising from the formation of $Ca(OH)_2$), which subsequently transforms to $CaCO_3$ (Slosarczyk et al., 1996). The presence of CaO may alter the rate and extent of biodegradation. A Ca/P less than 1.67 favours the presence of α-TCP or β-TCP in HA (Raynaud et al., 2002). The Ca/P ratio does not affect significantly the grain growth of HA (Suchanek and Yoshimura, 2011). Moreover, due to the formation of a new phase and dehydration, the decomposition has negative effect on densification of the HA.

Aoki (1991) reported elastic modulus of dense HA between 35 and 120 GPa (44–88 GPa in bending mode) which further depends on measurement technique adopted, porosity and impurity present. Strength of dense HA ceramics is in the range of 40–250 MPa (bending), 130–900 MPa (compressive) and 40–300 MPa (tensile), while Vickers hardness (H_v) is found in a range 3.0–7.0 GPa (Suchanek and Yoshimura, 1998). Wakai et al. (1990) reported that dense HA ceramics exhibit super-plasticity in a temperature range of 1000°C –1100°C due to sliding of grain boundary. Resistance to wear and coefficient of friction of dense HA are reported to be comparable (Rootare et al., 1978).

9.2.2.6 HA-Based Composites

With controlled physical and chemical parameters, dense as well as spongy HA ceramics can be produced. However, poor mechanical properties put the limitation on reliability of applications of HA. HA-based composites can be used to regulate the organic properties of implantations. Many reinforcement materials, including fibres, whiskers, platelets and particles are being used with HA to enhance its service life (de With and Corbijin, 1989, Juang and Hon, 1994, Nanomi and Satoh, 1995, Noma et al., 1993; Guo et al., 2003). The mechanical properties of HA composites will enhance at the cost of reduction in bioactivity and biocompatibility than that of pure HA, further the process related to production of HA composites may lead to the decomposition of HA to TCP and CaO phases.

Almost 40 years ago, in the late 1960s, first bioactive material (bio-glass) developed by Hench exhibits biocompatibility and bioactivity. The reinforcement of HA with bio-glass resulted in the improvement of mechanical properties without any reduction of biocompatibility or bioactivity. In such a composite, apatite crystalline phase crystallizes from glassy matrix during appropriate heat treatment. Bio-glass maintains high strength for longer time than HA both in vitro and in vivo (Kokubo, 1991). By the mid-1980s, bioceramic materials reached medical use in a variety of orthopaedic and dental applications (Hench and Polak, 2002). Different bioceramics which include Ca-P, Al_2O_3 and bio-glass are reviewed and classified with focus on bone attachment (Hench, 1991, Andrea, 2007). Researchers reported that despite improved strength and toughness, reinforced HA has lesser applications due to decrease in their bioactivity and biocompatibility and complications in processing.

9.2.2.7 HA-Based Composite Coatings

The metallic materials for body implants have very good mechanical properties but they have lack of bioactive and biocompatible, so their service life is restricted to 10–15 years and the failure is associated with the bone implant interface (Amaral, 2008), while the bioceramics have very good bioactivity and biocompatibility but

Biomaterials

they do not have even fair mechanical properties, where tensile strength, brittleness, fracture toughness and resistance to wear are concerned. Thus, the HA coatings on metallic implants forms a macro-composite material that can be used for clinical applications, which combines of strength of metallic substrate and compatibility of HA. Such composite provides natural fixation of insert to bone and reduces adverse effect due to release of metallic ions. Furthermore, HA coatings have been reported to promote early bone formation around implants and promote the deposition and differentiation of mesenchymal cells into osteoblasts on HA-coated implant surface (Cleries et al., 2000). Ability to osteoblast adhesion onto the surface is due to the increased adsorption and production of proteins. In addition, a study by Gottlander et al. (1992) has indicated that HA-coated implants have a higher percentage of bony contact after six weeks of implantation as opposite to cp titanium fastener. A predominately fibrous tissue interface was observed on titanium implant only with a small area of direct bone contact (Thomas et al., 1987). Gammage et al. (1990) reported that HA-coated implants showed an increased coronal bone growth than that observed with Ti implant in an animal study. Many investigators reported that bone adapts in much less time to HA-coated implants than to Ti implants (Gottlander and Albrektsson, 1991; Weimlaender et al., 1992).

The HA coatings were tested with respect to initial fixation, stable bone in-growth and remodelling around the stem in total hip prosthesis. Comparative studies have shown HA-coated stems had high life expectancy as compared to cemented one. Radiographic evaluation revealed a lesser ion migration, lower bone resorption and stable bone in-growth with HA coating (Johnson, 2005).

The HA-coated metallic implants are studied for many years. In spite of many advantages of HA, the brittle nature and low fracture toughness of HA coatings often result in rapid wear and premature failure of coatings. Hence, improvement in mechanical properties of HA coatings are desirable (Tercero et al., 2009). Although, HA-coated implants show strong bonding with bone, but stability of interface could be a problem either during or after implantation (Nie et al., 2000, Collier et al., 1993; Simmons et al., 1999).

The studies by various investigators suggested that to avoid this inherent deficiency, reinforcement of secondary phase with better mechanical properties may be enhanced. Various HA-based composites have been produced to solve this problem (Zheng et al., 2001, Hyuschenko et al., 2002; Chou and Chang, 2002a, 2002b).

Efforts are being made to enhance these properties of HA by reinforcing it with a secondary material such as Al_2O_3, ZrO_2, TiO_2, SiO_2, carbon nanotubes. Evis and Doremus (2005, 2007) studied HA-reinforced α-Al_2O_3 and HA/ZrO_2 composite coating and found increased hardness, better bonding and fracture toughness compared with pure HA coating.

Significant appreciation in mechanical properties and decrease in dissolution of plasma sprayed HA/10 wt% ZrO_2 composite coating as compared to HA coating was reported by Lee et al. (2004). However, no apatite was found on surface of HA/ ZrO_2 in vitro. A reduction in ZrO_2 contents of composite coating was suggested. Balamurugan et al. (2007) investigated the electrochemical behaviour of HA/ZrO_2-based sol-gel coating on surgical grade stainless steel and reported that the ZrO_2 regardless of its phase and composition did not altered HA matrix in the composite coating. Further, the reinforced ceramic possessed good in vitro bioactivity and bio-compatibility (Gu et al., 2003).

HA/15 wt%SiO_2/10wt% Ti was plasma sprayed on food grade stainless steel substrate (Morks, 2008). Adhesive tensile strength of coating was expected to be higher than that of cohesive (Morks et al., 2008). Results of investigation by Morks et al. (2008) confirmed the enhancement of the adhesive strength by the reinforcement of silicon oxide and titanium particles when compared to HA coatings. An impact theory was used to reasonably explain the coating of HA/TiO_2 composite and reaction of HA and titania during High Velocity Oxy-Fuel (HVOF) process by Li et al. (2003). It was found that the crater caused by impingement of completely unmelted titania particle upon HA matrix was much smaller than titania feedstock. A chemical bond of HA and titania was proposed, which may be responsible for the entrapment. Gu et al. (2003) reported in an investigation that plasma sprayed HA – 50wt% Ti-6Al-4V offered better properties than pure HA coatings leading to much better long-term stability of composite coatings in body environment. Zheng et al. (2000) reported the increase in bond strength to 14.5 MPa and 17.3 MPa from 12.9 MPa by the reinforcement of 20 and 60 wt% Ti, respectively. The plasma spray process was controlled to get titanium-rich coating near substrate surface and HA-rich coating at the top by Inagaki et al. (2006). The bond strength of coating prepared by plasma gas-containing Ar-N_2 varied from 1.5 to 2 times the bond strength of coating prepared by Ar-O_2 as plasma gas. Inagaki et al. (2003) examined the effect of partially nitriding the HA composite coating during plasma spraying and reported substantial increase in bonding strength by adding 0.8% N_2 to argon while spraying as compared to coatings sprayed by pure argon, further the bonding strength obtained was almost double by adding 1.8% N_2 to argon during plasma spraying as compared to pure argon as plasma gas.

Elastic modulus and toughness of HVOF-sprayed HA and HA-reinforced titania were examined by Li et al. (2002). The results showed substantial increase in elastic modulus by addition of 10 wt% titania to HA as compared to pure HA coating, while by addition of 20 wt% titania the value of modulus decreased but it was still higher than pure HA coating. The decrease in value of modulus by addition of 20 wt% titania is attributed to the significant enhanced influence of the multi-phases (Li et al., 2002). The fracture toughness of the coatings was increased due to increased content of Ti-O_2 (Hannora and Ataya, 2016). The chemical reaction of HA and titania leads to more dense structure and hence higher fracture toughness.

Lim et al. (1999) investigated HA/YSZ on Ti-6Al-4V as a function of plasma current by varying the current between 600 and 1000A. The results show that decomposition of HA was due to reaction of ZrO_2 with CaO (one of the phases present in plasma-sprayed HA coatings) (Lim et al., 1999). Bonding was improved from 28.6 to 32.5 MPa by addition of 10 wt% ZrO_2, while bond strength increased to 36.2 ± 3.0 with a ZrO_2 bond coat as reported by Chou and Chang (2002a, 2002b). The mechanism of increase in two-layer HA/ZrO_2 coating on rougher ZrO_2 intermediate coat, which promotes a better mechanical interlocking between HA top-coat and ZrO_2 intermediate coat was suggested (Chou and Chang 2002 a & b). Further, improved cohesive strength of HA coatings which contribute to bond strength of coating and substrate is due to reinforced ZrO_2 particles. Fu et al. (2001) prepared different powders based on HA with two different processes: HA-30 wt% YSZ ball milled and

Biomaterials

HA/30 wt% YSZ spheroidized powder and plasma spray coated on the substrates. The mechanical properties of spheroidized powder coating were found to be better than that of ball-milled powder coating (Dinarvand et al., 2011). There were more unmelted YSZ and HA particles in ball-milled powder coating due to irregular shape and agglomeration of particles (Fu et al., 2001).

Tercero et al. (2009) investigated the plasma-sprayed bioceramic composite coatings of HA-reinforced Al_2O_3 and CNT. The results showed an enhancement of fracture toughness by 158% by addition of 20 wt% alumina and 300% by addition of 18.4 wt% alumina and 1.6 wt% carbon nanotubes. The layering of Al_2O_3 on titanium-based alloy was suggested by Di Palma (2004) (Sobieszczyk, 2008). Alumina, which is classified as one of the chemically and biologically inert substance, has been widely investigated as a reinforcement material with HA (Champion et al., 1996).

Highly dense and pure alumina was the first material used medically as load-bearing hip prostheses and dental implant because of its combination of excellent strength and resistance to corrosion and wear, good biocompatibility and stability in physiological environment (Hulbert at el. 1987, Hulbert et al., 1987, Hench and Ethridge, 1982, Plenk, 1982; Griss and Heimke, 1981). Limited chemical bonding of alumina with tissue leads to its limited applications. In 2002, FDA approved Al_2O_3 articulated hips for marketing in the United States. Addition of nano-size alumina into HA resulted in smaller thermal expansion coefficient nearer to Ti-6Al-4V.

The interaction between HA and human monocites results in the liberation of inflammatory cytokines. The addition of 0.5%–2% Zn was proposed to retard the inflammatory processes (Grandjean-Laquerriere et al., 2006, Sobieszczyk, 2008).

Interlaminar shear strength, hardness and elastic modulus of monolithic HA was found to be increased by addition of 2% carbon nanotubes (CNT) deposited by electrophoresis deposition process (Kaya, 2008, Kwok, 2009).

9.2.2.8 Bond Coat

Bond or intermediate coat is a layer that is sprayed between substrate and top-coat. The addition of intermediate layer offers improvement in adhesion strength. The bond coat reduces the release of metallic ions from substrate, which improves biocompatibility in case of body implants. This layer of coating also reduces the thermal expansion mismatch at interface and hence reduces the stresses that give rise to cracking and delamination. The interface of HA and titanium is a site of critical weakness when compared with the strength of inter- and intralamellar structure of HA coating (Chou and Change 2002a). Investigators used different materials as bond coat including zirconia (ZrO_2), titania (TiO_2), alumina (Al_2O_3) and the combination of these oxides and dicalcium silicate.

The adhesive strength can be enhanced by mechanically interlocking interface (Tucker, 1982). Mechanical interlocking was established via porous anodic Al_2O_3 bond coat on physically vapour deposited (PVD) aluminium film on titanium by Wu et al. (2006).

Bonding strength of HA/ZrO_2 on Ti substrate increased substantially when intermediate layer of ZrO_2 was applied. Chou and Chang (2002a, 2002b) suggested mechanical interlocking between HA top-coat and ZrO_2 bond coat may be possible strengthening mechanism of two layer coating as rougher surface provided by ZrO_2

intermediate layer (Chou and Chang, 2002a, 2002b). The surface morphology of ZrO_2 bond coat promotes the bonding strength by increasing the surface area. In addition, the toughness of ZrO_2 may be a cause of strengthening (Chou, 2003). The interdiffusion of elements between HA and ZrO_2 also promotes a better interface bonding. Kurzweg et al. (1998 a & b) examined calcium oxide-stabilized zirconia ($CaOZrO_2$) and TiO_2+ZrO_2 intermediate layer. The enhancement of bonding at HA/ ZrO_2 interface due to diffusion of ions from HA top coat to ZrO_2 bond coat during deposition was examined via TEM.

Tomaszek et al. (2007) studied the behaviour of HA coatings with and without using TiO_2 as bond coating. Titanium oxide intermediate layer improves the adhesion and decreases the release of heavy ions. TiO_2 intermediate layer was reported to improve the quality of coating/titanium border (Tomaszek et al., 2007). Nie et al. (2000) reported that hybrid micro-arc discharge oxidation and electrophoresis-coated intermediate layer of HA/TiO_2 composite and top coat of HA considerably reduced the metal ion to human body and hence avoid deleterious effects. Kim et al. (2004) reported that affinity of TiO_2 with HA and Ti greatly contributed toward increased adhesion strength. The improvement in adhesion strength was found to as high as 60%. Kurzweg et al. (1998) confirmed that the bonding strength with titania bond coat was double the value of a HA coating. TiO_2 was deposited as bond coat and HA as top coat by electrophoresis process (Mohseni et al., 2014). The adhesion strength of coating increased with the decreasing voltage used for TiO_2 deposition from 21.0 MPa with 10 V to 11.9 MPa with 50 V. The adhesion strength of coating with 50 V was even lower than that of HA deposited with 200 V without intermediate layer (Albayrak et al., 2008). Kurzweg et al. (1998) reported that a 10–50-µm-thick dicalcium silicate bond coat increased the adhesion strength of HA coatings. A two-layer HA and HA/50 wt% titania bond coat composite coating on titanium substrates were fabricated by plasma spraying. By the introduction of bond coat, the residual stresses in top coat were found to decrease. The increase in toughness and strength of bond coat was due to TiO_2 as obstacles embarrassing cracking, stress-induced micro-cracking and decrease in coefficient of thermal expansion mismatch (Lu et al., 2004).

Bio-glass was sprayed on to titanium with and without Al_2O_3 40 wt% TiO_2 as bond coat by plasma spraying process. The bond strength of specimens with bond coat was increased by three times as compared with the specimens without bond coat and the adhesive bonding at interface turned into cohesive (Goller, 2004).

9.2.2.9 Significance of Bone Implant Interface

When attempt is made to regenerate bone via the conduction of bone into biomaterials, the canal is rooted adjoining to bone tissue. Implant when prosthesis in bone, it is mechanically fixed in place. Cells begin to occupy and populate the implant to form new bone by laying down new matrix.

The process of cell adhesion is directly linked to the type of tissue response at interface (Hench, 1991, Hench and Ethridge, 1982; De Aza, 2001). No implanted material in body system is passive; all materials produce a response from living tissues. The four types of response allow different means of achieving attachment of prostheses to musculoskeletal system.

Biomaterials 179

- If the implant material is toxic (Narayanan, 2008), the tissues surrounding the implant will die.
- If it is nontoxic and biologically inactive, a fibrous tissue of variable thickness will form.
- If the material is nontoxic but biologically active, an interfacial bond will form.
- If the material is nontoxic and dissolve, the surrounding tissue will replace the material.

The bone and coated implant interface are shown in Figure 9.4. There are two steps for the formation of fibrous capsule: first, the fixing of the anticipated cells, mediated by adsorbed proteins and second, the recruitment of osteogenic cells and their differentiation to osteoblasts by adequate cytokines.

On bare metallic implant, the bone will grow unilaterally toward the implant. When the bone trabeculae reach the implant's surface, they begin to spread parallel to the surface bridging the gap (Kettner, 1996). Many investigators reported that in case of HA-coated implants, bone can grow on both the surfaces i.e. on implant and bone thus filling the gap rapidly (Kettner, 1996, Porter et al., 2002; Soballe, 1996). The bidirectional gap filling allows fixation to occur almost double as quickly as it would for a bare metallic implant. The growth of bone cells on both uncoated titanium and HA-coated Ti insert is represented in Figure 9.5 that clearly shows the bidirectional filling on HA-coated implant.

The protection from metal ion release from metallic implant, the phenomenon which occurs at interface, is another advantage of HA-coated metallic substrate (Sun et al., 2002; Sousa and Barbosa, 1996). Release of metal ions causes the body to initiate an immune response by the formation of fibrous membrane around the implant, which prevents adequate fixation between the bone and the implant and reduces the load-bearing capacity of the joint. Soballe (1996) and Nagano et al. (1996)

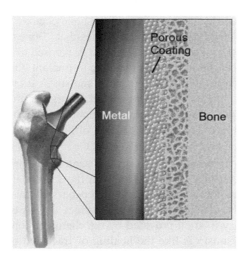

FIGURE 9.4 Interface of bone and HA-coated implant.

FIGURE 9.5 Micrographs showing osteointegration into HA-coated implant (Levingstone, 2008).

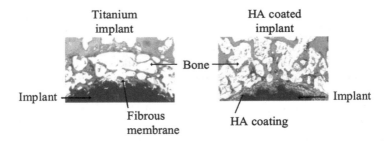

FIGURE 9.6 Micrographs presenting the formation of a fibrous membrane on titanium implant and no membrane on HA-coated titanium implant (Levingstone, 2008).

demonstrated that chemical composition of HA (which is similar chemical composition of bone) does not allow formation of fibrous membrane as demonstrated in Figure 9.6. The bone cells directly grow on the HA coating, a direct chemical bond can be formed between bone and coated implant, which allows the transfer of forces between the two more efficiently. Mechanical loading and transmission of forces have an important role in bone remodelling (Carlsson et al., 2004).

Loading is necessary for remodelling of bone; however, if the applied stresses produced are too less or too great, resorption of the bone occurs. The remodelling process is controlled by Wolff's law which postulates that "Bone continuously changes in order to cope with the mechanical loads that it is exposed to" (Currey et al., 1981). Other factors that affect the strength of the bone to implant bond includes the shape and topography of implant, the quality of bone and surgical factors, which includes quality of surgery techniques and factors related to surgical procedure

The dissolution/reprecipitation process is supposed to be the cause for the bone-bonding capability to HA. The process consists of fractional dissolution of HA and release of calcium and phosphate ions in form of Ca^{2+}, $H_2PO_4^-$, $H_2PO_4^{2-}$, PO_4^{3-} and $CaH_2PO_4^+$ in fluid surrounding the interface (Fazan and Marguis, 2000). Proteins and ions activate the surface of HA crystals on the surface of HA coating (Weng et al., 1997). Remodelling of spoiled bone also occurs in the conjunction with the coating dissolution. Further remodelling of interface occurs until a strong bond is formed. The secondary fixation provided by the chemical bond prevents loosening of implant. The mechanism is like the healing of fractured bone. Micromotion at the interface must be less than 50 µm in order to successful osseointegration and adequate fixation (Scheller and Jani, 2003).

Biomaterials

181

9.2.2.10 Comparison of Biological and Synthetic HA

Biological HA that is present in mineral contents of bone, teeth and enamel may contain many impurities, because the apatite structure is very friendly for substitution of many other ions. Biological HA is typically carbonate substituted and calcium deficient. The minor elements/ions associated with biological apatite are magnesium (Mg^{2+}), carbonate (CO_3^{2-}), sodium (Na^+), chloride (Cl^-), potassium (K^+), fluoride (F^-) and acid phosphate (HPO_4). Trace elements include strontium (Sr^{2+}), barium (Ba^{2+}) and lead (Pb^{2+}). The bioactivity and biocompatibility of synthetic HA is not only suggested by its similar composition to that of biological HA but also by the results of in vivo implantation, which has produced no inflammation, no foreign body response and no local or systemic toxicity (Klein et al., 1993). The confirmation of biocompatibility of HA is reported by Ducheyne et al. (1990), Ducheyne and Qiu (1999) and Buma et al. (1997). The composition of bone and synthetic HA are reported in Table 9.5.

9.3 FAILURE MECHANISM OF HA COATINGS

The failure mechanism of HA coatings is dissolution and wear. HA is slightly resolvable in water or acidic solution (Ryu et al., 2002). Several mechanisms are proposed by investigators to explain the dissolution behaviour of HA (Raemdonck et al., 1984; Le Geros, 1988). The pH value of less than 4 dramatically accelerates this process and $CaHPO_4$ (monetite) or $CaHPO_4 \cdot 2H_2O$ (DCPD, brushite) is more stable than HA at this pH value. There are three primary wear mechanisms, namely abrasive or third-body wear, adhesive wear and fatigue or delamination wear in bone and implant system.

Abrasive wear is the removal of material from one surface by other. Since no surface is perfectly smooth, local asperities on harder surface will plough through the softer material and gouge out wear particles. Presence of localized chemical bonding between two surfaces due to high pressure is the cause of adhesive wear. This wear is more common in hip replacement than in knee replacement. Fatigue or delamination wear causes the formation of sub-surface cracks and propagates due to cyclic

TABLE 9.5

Comparison of Natural (Found in Bone) and Synthetic HA Ceramic (Nicholson, 2002)

Composition (wt%)	Bone	HA
Ca	24.5	39.4
P	11.5	18.5
Ca/P ratio	1.65	1.67
Na	0.7	Traces
K	0.03	Traces
Mg	0.55	Traces
CO_3^{2-}	5.8	−

loading. The tip of crack experiences high stresses that propagate the crack until one crack joins the other and surface flakes off, which creates large wear particles. The third-body wear is most common in the bone implant system and will be studied in detail.

9.3.1 Dissolution Behaviour of Hydroxyapatite

The crystallinity and composition of calcium and phosphate phases and pH are the main factors which control the rate of in vitro or in vivo dissolution of HA. The dissolution rate also depends on the concentration and type of the surrounding solution, pH of solution, the degree of saturation, the solid-solution ratio and time of suspension in the solution. Factors such as the Ca/P ratio, impurities like F^- or Mg^{2+}, the degree of micro- and macro-porosities, defective structure and amount and type of other phases have significant effects on biodegradation. Two calcium phosphate materials are stable at room temperature in aqueous solution, and pH of the solution determines the stability of one of two materials (Klein et al., 1993). The solubility of various calcium phosphates in an aqueous solution at room temperature (25°C) is represented in Figure 9.7. In a thermodynamic analysis, the equilibrium diagram of predominant calcium and phosphorous is shown in Figure 9.8, while Table 9.6 gives all the equilibrium chemical reactions which could happen in Figure 9.8. At a pH lower than 4.2, dicalcium phosphate (DCP) is more stable, while at higher pH, greater than 4.2, HA is a stable phase (Schneider, 1991; Klein et al., 1993). The unhydrated high-temperature calcium phosphates when interacts with water or body fluid, at 37°C, forms HA.

$$4Ca_3(PO_4)_2 + 2H_2O \rightarrow Ca_{10}(PO_4)_6(OH)_2 + 2Ca^{2+} + 2HPO_4^{2-} \qquad (9.3)$$

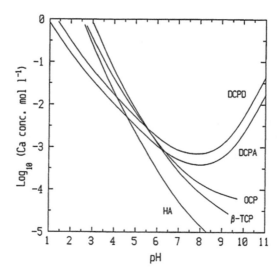

FIGURE 9.7 Solubility isotherms for various phases in the system $CaO-P_2O_5-H_2O$ at 25°C (Levingstone, 2008).

Biomaterials

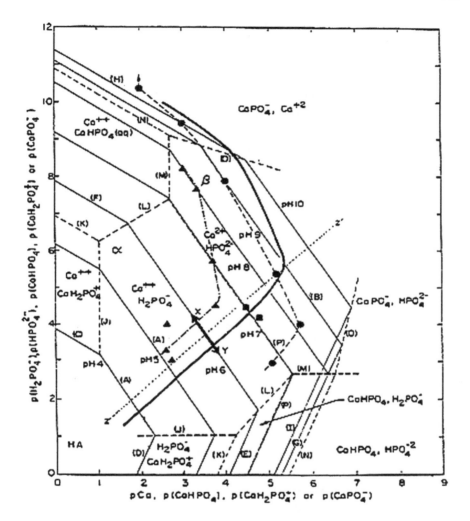

FIGURE 9.8 Equilibrium diagram for hydroxyapatite, solid lines are pH-invariant lines on the solubility surface and broken line separates the stability domain of various solution species (Chander et al., 1979).

The pH of physiological environment is 7.4. As presented in Figure 9.8, the crystalline HA is stable at these conditions, whereas β-TCP, octocalcium phosphate, DCP anhydrous and DCP dihydrate are less stable than crystalline HA at physiological conditions (Le Geros and Le Geros, 1993).

When HA is placed in an aqueous solution in the absence of any soluble salt of calcium or phosphorous, the system will equilibrate along dotted line Z-Z' as presented in Figure 9.8. Decomposition phases, such as calcium oxide (CaO), α-tricalcium phosphate (α-TCP), β-tricalcium phosphate (β-TCP), oxy-hydroxyapatite (OHA) and oxy-apatite (OA) are less stable than HA in vivo. The dissolution order that follows in the physiological solution is given (Heimann, 1999):

TABLE 9.6
Chemical Reactions in Ca-P-H$_2$O Solution System, Each of Reaction is Plotted in Figure 9.8 and Marked by Respective Alphabet (Chander et al., 1979)

A: $\quad Ca_{10}(PO_4)_6(OH)_2 + 14H^+ \leftrightarrow 10Ca^{2+} + 6H_2PO_4^- + 2H_2O$

B: $\quad Ca_{10}(PO_4)_6(OH)_2 + 8H^+ \leftrightarrow 10Ca^{2+} + 6HPO_4^{2-} + 2H_2O$

C: $\quad Ca_{10}(PO_4)_6(OH)_2 + 14H^+ \leftrightarrow 6CaHPO_4^+ + 4Ca^{2+} + 2H_2O$

D: $\quad Ca_{10}(PO_4)_6(OH)_2 + 4H_2PO_4^- + 14H^+ \leftrightarrow 10CaH_2PO_4^+ + 2H_2O$

E: $\quad Ca_{10}(PO_4)_6(OH)_2 + 4H_2PO_4^- + 4H^+ \leftrightarrow 10CaHPO_4(aq) + 2H_2O$

F: $\quad Ca_{10}(PO_4)_6(OH)_2 + 8H^+ \leftrightarrow 6CaHPO_4(aq) + 4Ca^{2+} + 2H_2O$

G: $\quad Ca_{10}(PO_4)_6(OH)_2 + 8H^+ + 4HPO_4^{2-} \leftrightarrow 10CaHPO_4 + 2H_2O$

H: $\quad Ca_{10}(PO_4)_6(OH)_2 + 2H^+ \leftrightarrow 4Ca^{2+} + 6CaPO_4^- + 2H_2O$

I: $\quad Ca_{10}(PO_4)_6(OH)_2 + 4HPO_4^{2-} \leftrightarrow 10CaPO_4^- + 2H^+ + 2H_2O$

J: $\quad CaH_2PO_4^+ \leftrightarrow Ca^{2+} + H_2PO_4$

K: $\quad CaH_2PO_4^+ \leftrightarrow CaHPO_4(aq) + H^+$

L: $\quad CaHPO_4(aq) + H^+ \leftrightarrow Ca^{2+} + H_2PO_4^-$

M: $\quad CaHPO_4(aq) \leftrightarrow Ca^{2+} + H_2PO_4^{2-}$

N: $\quad CaHPO_4(aq) \leftrightarrow CaPO_4^- + H^+$

O: $\quad CaPO_4^- + H^+ \leftrightarrow Ca^{2+} + HPO_4^{2-}$

P: $\quad H_2PO_4^- \leftrightarrow H^+ + HPO_4^{2-}$

$$CaO \gg ACP \gg \alpha-TCP \gg \beta-TCP \gg OHA/OA \gg HA \qquad (9.4)$$

Chow (1991) showed that the amount dissolved at equilibrium depends upon the thermodynamic solubility product of the compound and the pH of solution. Further, it was suggested that at pH of 7.0, the solubility decreases in the following order:

$$TTCP \gg \alpha-TCP \gg DCPD \gg DCPA \approx OCP \gg \beta-TCP \gg HA \quad (9.5)$$

Klein (1990) suggested that decomposition products formed even during the extremely short residence time of HA particles in hot plasma jet, dissolves more rapidly as compared to well-crystallized HA particles in the following order:

$$CaO \gg TTCO \gg ACP \gg TCP \gg HA \qquad (9.6)$$

The mechanism of degradation of calcium phosphate in the harsh body is not very clear. Some investigators, such as Nagano et al. (1996), de Groot (1984) and Yamada et al. (2001), observed that the process is a physio-chemical one in which particles are ingested by osteoclast cells attached to the surface and intracellular dissolution of these particles occurs.

Fazan and Marguis (2000) suggested that the dissolution process initiated at dislocations and grain boundary structures, while Porter et al. (2003) suggested that

Biomaterials

incoherent grain boundaries, without lattice continuity, are more sensitive to dissolution than semi-coherent grain boundaries. Bohner and Lemaitre (2009) from their findings suggested that the simplest simulated body fluid (SBF) in equilibrium with DCPD that mimics the main features of human blood plasma should be selected.

Amorphous and crystallize simultaneous vapour deposited coatings comprising of HA and α-TCP phases were tested in calcium-free Hank's solution by Hamdi and Ide-Ektessabi (2007) and reported that amorphous coatings almost immediately dissolved in a period of 1.5 days of immersion and α-TCP in crystallized coatings dissolved steadily with time resulting in the deterioration of coating surface. HA phase was found to be very stable and no sign of dissolution appeared even after the immersion period of 14 days. During the dissolution, reprecipitation of HA with preferred (211) orientation was observed in last days of immersion.

The unstable phase in HA coating which dissolves in physiological solution is undesirable as it leads to reduction in mechanical strength of coatings. In all the three coating, the OHA/α-TCP/β-TCP coatings performed better than the other two.

The HA coatings on SS 316L samples passivated at 0.64 V in borate buffer solution were found to exhibit better corrosion resistance in Ringer's solution, which serves as a simulated physiological medium. The better passivation behaviour is attributed to the formation of inner chromium oxide layer on the surface of stainless steel (Gopi et al., 2008). In another investigation, the implantation of nitrogen and helium ions was done on AISI 316L SS by Muthukumaran et al. (2010). A significant improvement in corrosion resistance and pitting corrosion was observed with both nitrogen and helium ion implantation as compared to virgin AISI 316L SS in NaCl solution (9 g/l of H_2O) at pH value of 6.3 at 37°C. Resistance to pit initiation, less susceptibility to localized attack and no change in HA phase during sintering are the findings of Kannan et al. (2003) on H_2SO_4-treated, HA-coated SS 316L in Ringer's solution (NaCl: 8.6g/l, $CaCl_2 \cdot 2H_2O$: 0.66 g/l and KCl: 0.6 g/l of H_2O). The presence of HA coating over passive layer was suggested as the primary cause, which played a dual role in preventing metal ion release and making the surface more bioactive.

Porter et al. (2002) compared in vivo behaviour of two HA coatings: first with a crystallinity of $70\% \pm 5\%$ and second an annealed coating with a crystallinity of $92\% \pm 1\%$. The non-annealed coatings demonstrated the precipitation of plate-like biological apatite crystallites adjacent to the coating after 3 hours. Similar bone growth type behaviour was not seen in the vicinity of the annealed coatings, which were more crystalline until a time period of 10 days.

The dissolved HA coating is resorbed by osteoclast cells. The resorption of coating is integrated into the normal bone remodelling process and is replaced with new bone with osteoblast cells. The dissolution rate of HA coatings is significant because a rapidly dissolving coating can lead to bone growth. However, this is always not the case. The rapid dissolution rate of HA coatings may lead to fast resorption, loss of fixation, implant loosening and production of particle debris (Hench and Wilson, 1993). Nevertheless, to obtain HA coatings with a predictable calcium phosphate phase, the deposition parameter must be optimized. In addition, the purity and crystallinity of the original HA powder must be controlled to prevent contamination. Highly crystalline and pure HA coatings generally result in the reduced healing time and faster fixation.

186 Functional and Smart Materials

9.3.2 UNRESOLVED ISSUES OF DEGRADATION OF HYDROXYAPATITE

Dissolution of HA is very important for remodelling of bone and bioactivity of HA-coated implant, it should be designed and controlled according to a specific clinical use. Due consideration should be given to the factors such as load-bearing and non-load-bearing sites, patient's sex and age, supply of blood in tissues, presence of cells in location of implant. To decide if the biodegradation is required or not, the form and composition of implant can be defined. However, due to limited knowledge regarding HA degradation, it is difficult to control the dissolution rate of HA. Some investigators showed that HA material failure in clinical use is related to HA dissolution (Kurashina et al., 2002). The HA biodegradation usually is not preferred in most of the cases, since a long-term mechanical strength of the implant is required. The fast dissolution of HA affects ceramic-induced osteogenesis and decreases bone ceramic bonding area (Kurashina et al., 2002). Although HA and TCP are composed of Ca, P, O and H elements, which are nontoxic to the body, but the debris of HA triggers the immune reaction, which results in the production of certain proteins. These proteins activate osteoclast and monocyte (Kim et al., 1993). After in vivo and in vitro examinations, microcracks and pits may be found in plasma-sprayed HA coating, which could cause mechanical failure of implant after long-term use (Gledhill et al., 2001b, Hemmerle et al., 1997; Bauer et al., 1993).

The inconsistent osseointegration of HA caused by HA dissolution is accompanied not only by biological conditions, but a diversity of HA material fabrication and use conditions also play an important role. For HA composite processing, it is more complicated according to the shape or form and performance which will be accomplished.

The other factors which influence the degradation of HA are impurities in HA, which broadly can be divided as compounds of calcium and phosphorous other than HA and foreign elements. The presence of foreign elements (Mg, C, F, Cl) is mainly from the raw materials. To control these impurities, quality of raw material and atmosphere during synthesis should be controlled, which are easy to control. The ASTM standard specification (ASTM F1185-88) (2003) states that ceramic HA for surgical implants has to have minimum HA contents of 95% by quantitative X-ray diffraction analysis. The concentration of heavy elements should be limited to the values shown in Table 9.7. The properties of HA required by the International Standards Institute (ISO 13778-1: 2000, Implants for surgery, Hydroxyapatite-Part I: Ceramic hydroxyapatite) (2000) should have a crystallinity of at least 45%, the maximum allowable

TABLE 9.7
Limits to Concentration of Heavy Metals

Elements	ppm (Maximum Limit)
Arsenic (As)	3
Cadmium (Cd)	5
Mercury (Hg)	5
Lead (Pb)	30

Biomaterials

limit of heavy metals up to 50 ppm and the Ca/P ratio for HA used for surgical implants between 1.65 and 1.82. The shape and microstructure of HA particles also affect the quality of coating.

REFERENCES

"Carbon Nanomaterials in Clean Energy Hydrogen Systems – II", Springer Science and Business Media LLC, 2011.

Albayrak, O., El-Atwani, O. and Altintas, S., (2008), "Hydroxyapatite coating on titanium substrate by electrophoretic deposition method: Effects of titanium dioxide inner layer on adhesion strength and hydroxyapatite decomposition", *Surf. Coat. Technol.*, Vol. 202, pp. 2482–2487.

Amaral, M., (2008), "Nanocrystalline diamond: In vitro biocompatibility assessment by MG63 and human bone marrow cells cultures", *J. Biomed. Mat. Res. Part A*, Vol. 87, pp. 91–99.

Andrea, S., (2007), "Drug delivery to the bone- implant interface: Functional hydroxyapatite surfaces and particles", Publikations server der Universität Regensburg.

Aoki, H., (1991), *"Science and Medical Application of Hydroxyapatite"*, Japanese Association of Apatite Science: Tokyo, Japan, pp. 179–192.

Balamurugan, A., Balossier, G., Kannan, S., Michel, J., Faure, J. and Rajeswari, S., (2007), "Electrochemical and structural characterization of zirconia reinforced hydroxyapatite bioceramic sol-gel coatings on surgical grade 316L SS for biomedical applications", *Ceram. Int.*, Vol. 33, pp. 605–614.

Bauer, T. W., Stulber, B. N., Ming, J. and Geesink, R. G. T., (1993), "Uncemented acetabular composites: Histologic analysis of retrieved hydroxyapatite-coated and porous implants", *J. Arthroplasty*, Vol. 8, pp. 167–177.

Bell, B. F. Jr., (2004), *"Functionally Graded, Multilayered Diamond Like Carbon-Hydroxyapatite Nanocomposite Coatings for Orthopedic Implants"*, M.Sc. Thesis, School of Materials Science and Engineering, Georgia Institute of Technology: Atlanta, GA.

Birtwistle, S. J., Wilson, K. and Porter, M. L., (1996), "Long-term survival analysis of total hip replacement", *Ann. Roy. Coll. Surg. England*, Vol. 78, pp. 180–183.

Black, J., (1999), *"Biological Performance of Materials: Fundamentals of Biocompatibility"*, Marcel Dekker Inc.: Basel.

Bohner, M. and Lemaitre, J., (2009), "Can bioactivity be tested *in vitro* with SBF solution?", *Biomaterials*, Vol. 30, pp. 2175–2179.

Bonfield, W., (1987), "New trends in implant materials", In *Biomaterials in Clinical Applications, Proc. of 6th European Conference on Biomaterials*, Eds. Pizzoferrato, A., Elsevier Science Publishers, New York, USA pp. 13–21.

Boskey, A. L. and Posner, A. S., (1973), "Conversion of amorphous calcium phosphate to microcrystalline hydroxyapatite. A pH dependent, solution mediated, solid-solid conversion", *J. Phys. Chem.*, Vol. 77, pp. 2313–2317.

Browne, M and Gregson, P. J., (1995), "Metal ion release from wear particles produced by Ti-6Al-4V and Co-Cr alloy surfaces articulating against bone", *Mater. Lett.*, Vol. 24, pp. 1–6.

Buma, P., Van Loon, P. J. M., Versleyen, H., Weinams, H., Slooff, T. J. J. H., de Groot, K. and Huiskes, R., (1997), "Histological and biomechanical analysis of bone and interface reactions around hydroxyapatite-coated intramedullary implants of different stiffness: A pilot study on the goat", *Biomaterials*, Vol. 18, pp. 1251–1260.

Carlsson, L. V., MacDonald, W., Jacobsson, C. M. and Albrektsson, T., (2004), "Osseointegration principles in orthopedics: Basic research and clinical applications", In *Biomaterials in Orthopedics*, Eds. Yaszemski, M. J., Trantolo, D. J., Lewandrowski, K., Hasirci, V., Altobelli, D. E. and Wise D. L., Marcel Dekker: New York, pp. 223–239.

Champion, E., Gautier, S. and Bernache-Assollant, D., (1996), "Characterization of hot pressed Al_2O_3-platelet reinforced hydroxyapatite composites", *J. Mater. Sci. Mater. Med.*, Vol. 7, pp. 125–130.

Chander, S. and Fuerstenau, D. W., (1979), "Interfacial properties and equlibria in apatite-aqueous solution system", *J. Colloid Interface Sci.*, Vol. 70, (3), pp. 506–516.

Chou, B-Y. and Chang, E., (2002a), "Plasma-sprayed hydroxyapatite coating on titanium alloy with ZrO_2 second phase and ZrO_2 intermediate layer", *Surf. Coat. Technol.*, Vol. 153 (1), pp. 84–92.

Chou, B-Y. and Chang, E., (2002b), "Plasma-sprayed zirconia bond coat as an intermediate layer for hydroxyapatite coating on titanium alloy substrate", *J Mater. Sci. Mater. Med.*, Vol. 13, pp. 589–595.

Chou, B.-Y., (2003), "Influence of deposition temperature on mechanical properties of plasma-sprayed hydroxyapatite coating on titanium alloy with ZrO_2 intermediate layer", *J. Thermal Spray Technol.*

Chow, L. C., (1991), "Development of self-setting calcium phosphate cements", *J. Ceram. Soc. Jpn. (Int. Ed.)*, Vol. 99, pp. 927–935

Chow, L. C., (2000), "Calcium phosphate cements: Chemistry, properties and applications", In *Proc. Materials Research Society*, Vol. 599, pp. 27–37.

Cleries, L., Martinez, E. and Fernandez Pradas, J. M., (2000), "Mechanical properties of calcium phosphate coatings deposited by laser ablation", *Biomaterials*, Vol. 21, pp. 967–971.

Collier, J. P., Suprenant, V. A., Mayor, M. B., Wrona, M., Jensen, R. E. and Suprenant, H. P., (1993), "Loss of hydroxyapatite coating on retrieved total hip components", *J. Arthroplasty*, Vol. 8, pp. 389–392.

Currey, J., Unsworth, A. and Hall, D. A., (1981), "Properties of bone, cartilage and synovial fluid", In *An Introduction to the Bio-mechanics of Joints and Joint Replacement*, Eds. Dowson, D. and Wright, V., Mechanical Engineering Publications Ltd.: London, pp. 103–119.

De Aza, P. N., (2001), "Transmission electron microscopy of the interface between bone and pseudo wollastonite implant", *J. Microscopy*, Vol. 201, pp. 33–43.

De Groot, K., (1984), "Calcium phosphate ceramics: Their current status", In *Contemporary Biomaterials-Material and Host Response, Clinical Applications, New Technology and Legal Aspects*, Eds., Boretos, J. W. and Eden, M., Noyes Publications: Park Ridge, NJ, pp. 477–492.

De With, G. and Corbjin, A. J., (1989), "Metal Fiber reinforced hydroxyapatite ceramics", *J. Mater. Sci.*, Vol. 24, pp. 3411–3415.

Di Palma, J.A., (2004), "Current treatment options for chronic constipation" *Rev Gastroenterological Disorders*, Vol. 4, pp. S34–S42.

Dinarvand, P., Seyedjafari, E. Shafiee, A., Jandaghi, A. B. et al., (2011), "New approach to bone tissue engineering: Simultaneous application of hydroxyapatite and bioactive glass coated on a poly(-lactic acid) scaffold", *ACS Appl. Mater. Interf*, Vol. 3, pp. 4518–4524.

Dorozhkin, S. V., and Epple, M., (2002) "Biological and medical significance of calcium phosphates", *Angewandte Chemie Int. Edition*, Vol. 41, pp. 3130–3146.

Ducheyne, P., Beight, J., Cuckler, B., Evans, B. and Radin, S., (1990a), "Effect of calcium phosphate coating characteristics on early post-operative bone tissue ingrowth", *Biomaterials*, Vol. 11, pp. 531–540.

Ducheyne, P. and Qiu, Q., (1999), "Bioactive ceramics: The effect of surface reactivity on bone formation and bone cell function", *Biomaterials*, Vol. 20, pp. 2287–2303.

Emery, D. F. G., Clarke, H. J. and Grover, M. L., (1997), "Stanmore total hip replacement in younger patients - review of a group of patients under 50 years of age at operation", *J. Bone Joint Surg. (Br)*, Vol. 79B, pp. 240–246.

Evis, Z. and Doremus, R. H., (2005), "Coatings of hydroxyapatite – nanosize alpha alumina composite on Ti-6Al-4V", *Mater. Lett.*, Vol. 59 (29–30), pp. 3824–3827.

Biomaterials

Evis, Z. and Doremus, R. H., (2007), "Hot pressed hydroxylapatite/monoclinic zirconia composites with improved mechanical properties", *J. Mater. Sci.* Vol. 42, pp. 2426–2431.

Fazan, F. and Marguis, P. M., (2000), "Dissolution behavior of plasma-sprayed hydroxyapatite coatings", *J. Mater. Sci. Mater. Med.*, Vol. 11, pp. 787–792.

Fu, L., Khor, K.A. and Lim, P. J., (2001), "The evaluation of powder processing on microstructure and mechanical properties of hydroxyapatite (HA)/yttria stabilized zirconia (ysz) composite coatings", *Surf. Coat. Technol.*, Vol. 140, pp. 263–268.

Gammage, D. D., Bowman, A. E., Meffert, R. M., Cassingham, R. J. and Davenport, W. A., (1990), "A histologic and scanning electron micrographic comparison of the osseous interface in loaded IMZ and Integral implants", *Int. J. Periodontic Restorative Dentistry*, Vol. 10, pp. 125–135.

Gledhill, H. C., Turner, I. G. and Doyle, C., (2001B), "In vitro dissolution behavior of two morphologically different thermally sprayed hydroxyapatite coatings", *Biomaterials*, Vol. 22, pp. 695–700.

Goller, G., (2004), "The effect of bond coat on mechanical properties of plasma sprayed bioglass-titanium coatings", *Ceram. Int.*, Vol. 30, pp. 351–355.

Gopi, D., Collins, A. P. V. and Kavitha, L., (2008), "Evaluation of hydroxyapatite coatings on borate passivated 316L SS in ringer's solution", *Mater. Sci. Eng. C*, Vol. 29 (3), pp. 955–958.

Gottlander, M. and Albrektsson, T., (1991), "Histomorphometric studies of hydroxyapatite coated and uncoated CP titanium threaded implants in bone", *Inter. J. Oral Maxillofac Implants*, Vol. 6, pp. 399–404.

Gottlander, M., Albrektsson, T. and Carlsson, L. V., (1992), "Histomorphometric studies of hydroxyapatite coated and uncoated CP titanium threaded implants in bone", *Inter. J. Oral Maxillofac Implants*, Vol. 7, pp. 485–490.

Gou, K. W., (2015), *"Surface Engineered Nanostructures on Metallic Biomedical Materials for Anti-Abrasion"*, Elsevier BV.

Grandjean-Laquerriere, A., Laquerriere, P., Jallot, E., Nedelec J. M., Guenounou, M., Laurent-Maquuin, D. and Philips, T. M., (2006), "Influence of zinc concentration of sol-gel derived zone substituted hydroxyapatite on cytokine production by human monocytes *in vitro*", *Biomaterials*, Vol. 27, pp. 3195–3200.

Griss, P. and Heimke, G., (1981), "Five years experience with ceramic-metal composite hip endo-prostheses", *Arch. Orthop. Traumat. Surg.*, Vol. 98, pp. 157–165.

Gu, Y. W., Khor, K. A. and Cheang, P., (2003), *"In vitro* studies of plasma-sprayed hydroxy-apatite/Ti-6Al-4V composite coatings in simulated body fluid (SBF)", *Biomaterials*, Vol. 24, pp. 1603–1611.

Guo, H., Khor, K. A., Boey, Y. C. and Miao, X., (2003), "Laminated and functionally graded hydroxyapatite/yttria stabilized tetragonal zirconia composites fabricated by spark plasma sintering", *Biomaterials*, Vol. 24, pp. 667–675.

Hamdi, M. and Ide-Ektessabi, A., (2007), "dissolution behavior of simultaneous vapor deposited calcium phosphate coatings in vitro", *Mater. Sci. Eng. C*, Vol. 27, pp. 670–674.

Hannora, A. E., and Ataya, S., (2016), "Structure and compression strength of hydroxyapatite/titania nanocomposites formed by high energy ball milling", *J. Alloys Comp*, Vol. 658, pp. 222–223.

Heimann, R. B., (1999), "Design of novel plasma sprayed hydroxyapatite-bond coat bioceramic systems", *J. Therm. Spray Technol.*, Vol. 8 (4), pp. 597–604.

Hemmerle, J., Qncag, A. and Erturk, S., (1997), "Ultrastructural features of bone response to a plasma-sprayed hydroxyapatite coating in sheep", *J. Biomed. Mater. Res.*, Vol. 36, pp. 418–425.

Hench, L. L. and Ethridge, E. C., (1982), *"Biomaterials: An interfacial Approach"*, Academic Press: New York.

Hench, L. L., (1991), "Bioceramics: From concept to clinic", *J. Am. Ceram. Soc.*, Vol. 74, pp. 1487–1510.

Hench, L. L. and Wilson, J., (1993), "An introduction to bioceramic", In *Advanced Series in Ceramics*, World Scientific, Singapore, Vol. 1, pp. 1–23.

Hench, L. L., (1998), "Bioactive materials: The potential for tissue regeneration", *J. Biomed. Mat. Res.*, Vol. 41 (4), pp. 511–518.

Hench, L. L. and Polak, J. M., (2002), "Third-generation biomedical materials", *Science*, Vol. 295, pp. 1014–1017.

Holt, C. l., (2009), "Role of calcium phosphate nanoclusters in the control of calcification", *FEBS J*, Vol. 276, pp. 2308–2323.

Hughes, J. M. and Rakovan, J., (2002), "The crystal structure of apatite, Ca5(PO4)3 (F, OH, Cl)", In *Phosphates: Geochemical, Geobiological, and Materials Importance, Reviews in Mineralogy and Geochemistry*, Eds., Kohn, M. J., Rakovan, J. and Hughes, M., ISBN 093996060X, Vol. 48.

Hulbert, S. F., Bokros, J.C., Hench, L. L., Wilson J. and Heimke, G., (1987), "Ceramics in clinical applications: Past, present and future", In *High Technology Ceramics*, Eds. Vineenzini, P., Elsevier: Amsterdam, pp. 189–213.

Hyuschenko, A. P., Okovity, V. A. and Shevtsov, A. I., (2002), " Investigation of composite hydroxyapatite powder for plasma spraying bioceramic coatings", *Mater. Manuf. Processes,* Vol. 17 (2), pp. 177–185.

Inagaki, M., Yokogawa, Y. and Kameyama, T., (2003), "Bond strength improvement of hydroxyapatite/titanium composite coating by partial nitriding during RF-thermal plasma spraying", *Surf. Coat. Technol.*, Vol. 173, pp. 1–8.

Inagaki, M., Yokogawa, Y. and Kameyama, T., (2006), "Effects of plasma gas composition on bond strength of hydroxyapatite/titanium composite coatings prepared by RF-plasma spraying", *J. Eur. Ceram. Soc.*, Vol. 26, pp. 495–499.

Jackson, M. J., (2006) Application nanotechnology, Surface Engineering, *Proceedings of the 5th international Surface Engineering Congress*, Washington, DC.

Johnson, S., (2005), *"Pulsed Laser Deposition of Hydroxyapatite Thin Films"*, M.Sc. Thesis, School of Materials Science and Engineering, Georgia Institute of Technology: Atlanta, GA.

Juang, H. Y. and Hon, M. H., (1994), "Fabrication and mechanical properties of hydroxyapatite-alumina composites", *Mater. Sci. Eng. C*, Vol. 2, pp. 77–81.

Kannan, S., Balamurugan, A. and Rajeswari, S., (2003), "Hydroxyapatite coatings on sulfuric acid treated type 316L SS and its electrochemical behavior in ringer's solution", *Mater. Lett.*, Vol. 57, pp. 2382–2389.

Kaya, C., (2008), "Electrophoretic deposition of carbon nanotubes-reinforced hydroxyapatite bioactive layers on Ti-6Al-4V alloys for biomedical applications", *Ceram. Int.*, Vol. 34, pp. 1843–1847.

Kettner, R., (1996), "The unique interface reaction of H-A-C", In *Proceedings of a Two Day Symposium on Hydroxyapatite Ceramics, A Decade of Experience in Hip Arthroplasty*, Eds. Furlong, R., Royal College of Surgeons of England: London, 2–3 November, Furlong Research Foundation, pp. 41–55.

Kim, K. J., Sato, K., Kotabe, S., Katoh, Y. and Itoh, T., (1993), "Biochemical and histochemical analysis of bone marrow cells activated by HA particles", *Bioceramics*, Vol. 6, pp. 365–369.

Kim, H., Koh, Y., Li, L., Lee, S. and Kim, H., (2004), "Hydroxyapatite coating on titanium substrate with titania buffer layer processed by sol-gel method", *Biomaterials*, Vol. 25, pp. 2533–2538.

Klein, C. P. A. T., (1990), "Study of solubility of different calcium phosphates and other ceramic particles in vitro", *Biomaterials*, Vol. 11, pp. 509–512.

Biomaterials 191

Klein, C. P. A. T., Wolke, J. G. C. and de Groot, K., (1993), "Stability of calcium phosphate ceramics and plasma spray coating", In *An Introduction to Bioceramic*, Eds. Hench, L. L. and Wilson, J., World Scientific: London, pp. 192–221.

Kokubo, T., (1991), "Bioactive glass ceramics: Properties and applications", *Biomaterials*, Vol. 12, pp. 155–163.

Kumari, R. and Dutta, J. M., (2017), "Microstructure and surface mechanical properties of plasma spray deposited and post spray heat treated hydroxyapatite (HA) based composite coating on titanium alloy (Ti-6Al-4V) substrate", *Materials Characterisation*.

Kurashina, K., Kurita, H., Wu, Q., Ohtsuka, A. and Kobayashi, H., (2002), "Ectopic osteogenesis with biphaic ceramics of hydroxyapatite and tricalcium phosphate in rabbits", *Biomaterials*, Vol. 23, pp. 407–412.

Kurzweg, H., Heimann, R. B. and Troczynski, T., (1998a), "Adhesion of thermally sprayed hydroxyapatite-bond-coat systems measured by a novel peel test", *J. Mater. Sci. Mater. Med.*, Vol. 9, pp. 9–16.

Kurzweg, H., Heimann, R. B., Troczynski, T. and Wayman, M. L., (1998b), "Development of plasma-sprayed bioceramic coatings with bond coats based on titania and zirconia", *Biomaterials*, Vol. 19, pp. 1507–1511.

Kwok, C.T., (2009), "Characterization and corrosion behavior of hydroxyapatite coatings on Ti6Al4V fabricated by electrophoretic deposition", *Applied Surface Science*.

Le Geros, R. Z., (1988), "Significance of porosity and physical chemistry of calcium phosphate ceramics, biodegradation-bioresorption", In *Bioceramics: Material Characteristics Versus In Vivo Behavior*, Eds., Ducheyne, P. and Lemons, J. E., Ann. N. Y. Acad. Sci., New york, Vol. 523, pp. 268–271.

Le Geros, R. Z., (1991), "Calcium phosphate in oral biology and medicine", In *Monographs in Oral Science*, Eds. Myers, H., Karger: Basel, Vol. 15, pp. 1–201.

Le Geros, R. Z. and Le Geros, J. P., (1993), "Dense hydroxyapatite", In *An Introduction to Bioceramics*, Eds., Hench, L. L. and Wilson, J., World Scientific: London, pp. 139–180.

Le Geros, R. Z., (1994), "Biological and synthetic apatites", In *Hydroxyapatite and Related Materials*, Eds. Brown, P. W. and Constantz, B., CRC Press: Boca Raton, FL, pp. 3–28.

Lee, T. M., Yang, C. Y., Chang, E. and Tsai, R. S., (2004), "Comparison of plasma-sprayed hydroxyapatite coatings and zirconia-reinforced hydroxyapatite composite coatings: In vivo study", *J. Biomed. Mater. Res. A*, Vol. 71A, pp. 652–660

Levingstone, T. J., (2008), *"Optimisation of Plasma Sprayed Hydroxyapatite Coatings"*, Ph. D. Thesis, School of Mechanical and Manufacturing Engineering, Dublin City University: Dublin.

Li, H., Khor, K. A. and Cheang, P., (2002), "Young's modulus and fracture toughness determination of high velocity oxy-fuel-sprayed bioceramic coatings", *Surf. Coat. Technol.*, Vol. 155, pp. 21–32.

Li, H., (2003), "Impact formation and microstructure characterization of thermal sprayed hydroxyapatite/titania composite coatings", *Biomaterials*.

Li, H., Khor, K. A. and Cheang, P., (2003), "Impact formation and microstructure characterization of thermal sprayed hydroxyapatite/titania composite coatings", *Biomaterials*, Vol. 24, pp. 949–957.

Lim, V. J. P., Khor, K. A., Fu, L. and Cheang, P., (1999), "Hydroxyapatite-zirconia composite coatings via the plasma spraying process", *J. Mater. Proces. Technol.*, Vol. 89–90, pp. 491–496.

Liven, L., (2007), *"Study of Hydroxyapatite Osteoinductivity with an Osteogenic Differentiation Assay using Mesenchymal Stem Cell"* M. Phil. Thesis, Hong Kong University of Science and Technology: Hong Kong.

Long, M. and Rack, H. J., (1998), "Titanium alloys in total joint replacement – A materials science perspective", *Biomaterials*, Vol. 19, pp. 1621–1639.

Lu, Y-P., Li, M-S., Li, S-T., Wang, Z-G. and Zhu, R-F., (2004), "Plasma-sprayed hydroxy-apatite + titania composite bond coat for hydroxyapatite coating on titanium substrate", *Biomaterials*, Vol. 25, pp. 4393–4403.

Mann, S., (2001), *"Biomineralization-Principles and Concepts in Bioinorganic Materials Chemistry,"* Oxford University Press: Oxford.

Mittal, M., Nath, S.K. and Prakash, S., (2013), "Improvement in mechanical properties of plasma sprayed hydroxyapatite coatings by Al2O3 reinforcement", *Mater. Sci. Eng.: C.* Mohseni, E., Zalnezhad, E. and Bushroa, A.R., (2014), "Comparative investigation on the adhesion of hydroxyapatite coating on Ti–6Al–4V implant: A review paper", *Int. J. Adhes. Adhes* Vol. 48, pp. 238–257.

Morks, M. F., (2008), "Fabrication and characterization of plasma sprayed HA/SiO_2 coatings for biomedical application", *J. Mech. Behav. Biomed. Mater.*, Vol. 1, pp. 105–111.

Morks, M.F., Fahim, N.F. and Kobayashi, A., (2008), "Structure, mechanical performance and electrochemical characterization of plasma sprayed SiO2/Ti-reinforced hydroxy-apatite biomedical coatings", *Appl. Surf. Sci.*, Vol. 255, pp. 3426–3433.

Muthukumaran, V., Selladurai, V., Nandhakumar, S. and Senthilkumar, M, (2010), "Experimental investigation on corrosion and hardness of ion implanted AISI 316L stainless steel", *Mater. Des.*, Vol. 31, pp. 2813–2817.

Nagano, M., Nakamura, T., Kokubo, T., Tanahashi, M. and Ogawa, M., (1996), "Differences of bone bonding ability and degradation behavior in vivo between amorphous calcium phosphate and highly crystalline hydroxyapatite coating", *Biomaterials*, Vol. 17 (9), pp. 1771–1777.

Nanomi, T. and Satoh, N., (1995), "Preparation of elongated diopside/hydroxyapatite compos-ite and their cell culture test", *J. Ceram. Soc. Japan*, Vol. 103, pp. 804–809.

Narayanan, R., (2008), "Calcium phosphate-based coatings on titanium and its alloys", *J. Biomed. Mater. Res. B Appl. Biomater.*, Vol. 85, pp. 279–299.

Netter, F. H., Hammoud, G., Kozinn, H. C., Wilson, P. D., Jr. and Pillicci, P. M., (1989), "The ciba collection of medical illustrations", In *Musculoskeletal System, Part. II*, Eds., Summit, N. J., Ciba Geigy Corporation Switzerland, Vol. 8, pp. 235–246.

Nicholson, J. W., (2002), *"The Chemistry of Medical and Dental Materials"*, Royal Society of Chemistry: Cambridge.

Nie, X., Leyland, A. and Mattews, A., (2000), "Deposition of layered bioceramic hydroxyapa-tite/TiO_2 coatings on titanium alloys using a hybrid technique of micro-arc oxidation and electrophoresis", *Surf. Coat. Technol.*, Vol. 125, pp. 407–414.

Noma, T., Shoji, N., Wada, S. and Suzuki, T., (1993), "Preparation of spherical Al_2O_3 particle dis-persed hydroxyapatite ceramics", *J. Ceram. Soc. Japan, (Int. Ed.)*, Vol. 101, pp. 923–927.

Park, J. B., and Lakes, R. S., (2007), *"Biomaterials an Introduction"*, 3rd Ed., Springer, New York, USA, ISBN:978-0-387-37879-4.

Plenk, H., (1982), "Biocompatibility of ceramics in joint prostheses", In *Biocompatibility of Orthopedic Implants*, , Vol. 1, pp. 269–295, CRC Press: Boca Raton, FL.

Porter, A. E., Hobbs, L. W., Rosen, V. B. and Spector, M., (2002), "The Ultrastructure of plasma-sprayed hydroxyapatite-bone interface predisposing to bone bonding", *Biomaterials*, Vol. 23 (2), pp. 725–733.

Porter, A. E., Patel, N., Skepper, J. N., Best, S. M. and Bonfield, W., (2003), "Comparison of in vivo dissolution processes in hydroxyapatite and silicon-substituted hydroxyapatite bioceramic", *Biomaterials*, Vol. 24, pp. 4609–4620.

Raemdonck, W. V., Ducheyne, P. and DeMeestar, P., (1984), "Calcium phosphate ceramics", In *Metal and Ceramic Biomaterials II-Strength and Surface*, Eds., Bucheyne, P. and Hastings, G. W., CRC Press, Boca Raton, FL, pp. 143–166.

Raynaud, S., Champion, E. and Bernache-Assollant, D., (2002), "Calcium phosphate apatites with variable Ca/P atomic ratio, calcination and sintering", *Biomaterials*, Vol. 23, pp. 1073.

Biomaterials

Rootare, H. M., Powers, J. M. and Craig, R. G., (1978), "Sintered hydroxyapatite ceramic for wear studies", *J. Dent. Res.* Vol. 57, pp. 777–783.

Ryu, S., Youn, H. J., Hong, K. S., Kim, S. J., Lee, D. H. and Chang, B. S., (2002), "Correlation between MgO doping and sintering characteristics in hydroxyapatite/β-tricalcium phosphate composite", *Key Eng. Mater.*, Vol. 21–24, pp. 218–220.

Scheller, G. and Jani, L., (2003), "The cementless total hip arthroplasty", In *Pelvic Ring and Hip*, Eds. DuParc, J., Elsevier: Paris, Vol. 6.

Schneider, S. J. Jr., (1991), *"Ceramics and Glasses"*, ASM international: Cleveland, OH Vol. 4.

Sedlak, J. M. and Beebe, R. A., (1974), "Temperature programmed dehydration of amorphous calcium phosphate", *J. Colloid. Interface Sci.*, Vol. 47, p. 483.

Shi, J., (2004), *"From Dental Enamel to Synthetic Hydroxyapatite-based Biomaterials"*, Ph.D. Thesis, Department of Earth Sciences, University of Hamburg: Hamburg.

Simmons, C. A., Valiquette, N. and Pilliar, R. M., (1999), "Osseointegration of sintered porour-surfaced and plasma spray-coated implants: An animal model study of early post implantation healing response and mechanical stability", *J. Biomed. Mater. Res.*, Vol. 47 (2), pp. 127–138.

Slosarczyk, A., Stobierska, E., Paszkiewicz, Z. and Gawlicki, M., (1996), "Calcium phosphate materials prepared from precipitates with various calcium: Phosphorus molar ratio", *J Am. Ceram. Soc.*, Vol. 79. p. 2539.

Soballe, K., (1996), "The role of H-A-C in in-growth prosthesis", In *Proceedings of Two Day Symposium on Hydroxyapatite Ceramics, A Decade of Experience in Hip Arthroplasty*, Eds. Furlong, R., Furlong Research Foundation, Royal College of Surgeons of England, London, 2nd -3rd November, pp. 57–67.

Sobieszczyk. S., (2008) "Coatings in arthroplasty: Review paper", *Adv. Mater. Sci.*, Vol 8, pp. 35–54.

Sousa, S. R. and Barbosa, M. A., (1996), "Effect of hydroxyapatite thickness on metal ion release from Ti6Al4V substrates", *Biomaterials*, Vol. 17, pp. 397–404.

Spector, M., (1990), "Prostheses: Materials, design and strategies for fixation in orthopedic knowledge update", *Am. Acad. Orthopedic Surgeons*, Vol. 3, pp. 115–129.

Spector, M., Shortkroff, S., Sldge, C. B. and Thornhill, T.S., (1991), "Advances in our understanding in implant-bone interface: Factors affecting formation and degradation", In *Instructional Course Lecture XI*, Eds. Tullos, H.S., American Academy of Orthopedic Surgeons: Park Ridge, IL, pp. 101–113

Sprio, S., (1970), *"Synthesis and Characterization of Implants for Bone Substitutions Made of Biomedical Apatites Containing Silicon"*, Alma Mater Studiorum - Università di Bologna: Bologna.

Suchanek, W. and Yoshimura, M., (1998), "Processing and properties of hydroxyapatite-based biomaterials for use as hard tissue replacement implants", *J. Mater. Res.* Vol. 13 (1), pp. 94–117.

Sun, L., Berndt, C. C., Khor, K. A., Gross, K. A. and Cheang, H. N., (2002), "Surface characterization and dissolution behavior of plasma-sprayed hydroxyapatite coating", *J. Biomed. Mater. Res.*, Vol. 62, pp. 228–236.

Tercero, J. E., Namin, S., Lahiri, D., Balani, K., Tsoukyas, N. and Agarwal, A., (2009), "Effect of carbon nanotube and aluminum oxide addition on plasma-sprayed hydroxyapatite coating's mechanical properties and biocompatibility", *Mater. Sci. Eng., C*, Vol. 29, pp. 2195–2202.

Thomas, K. A., Kay, J. F., Cook, S. D. and Jarcho, M., (1987), "The effect of surface macro-texture and hydroxyapatite coating on the mechanical strength and histologic profiles of titanium implant material", *J. Biomed. Mater. Res.*, Vol. 21, pp. 1395–1414.

Todd, R. C., Lightowl, C. D. and Harris, J., (1972), "Total hip replacement in osteioarthrosis using charnley prosthesis", *Br. Med. J.*, Vol. 2, p. 752.

Tomaszek, R., Pawlowwski, L., Gengembre, L., Laureyns, J. and le Maguer, A., (2007), "Microstructure of suspension plasma sprayed multilayer coatings of hydroxyapatite and titanium dioxide", *Surf. Coat. Technol.*, Vol. 201, pp. 7432–7440.

Tucker, C., (1982), *"Hand-Book of Deposition Technologies for Films and Coatings"*, Eds. Bunshah, R.F., Noyes Publications: Park Ridge, NJ, ISBN: 0–8155-1337-2 pp. 617–665.

Vallet-Regi, M. and Gonzalez-Calbet, J. M., (2004), "Calcium phosphate as substitution of bone tissues", *Prog. Solid State Chem.*, Vol. 32 (1–20), pp. 1–31.

Wakai, F., Kodama, Y., Sakagawa, S. and Nonami, T., (1990), "Superplasticity of hot isostatically pressed hydroxyapatite", *J. Am. Ceram. Soc.*, Vol. 73, pp. 457–460.

Weast, R.C., (1985–1986), "The CRC handbook of chemistry and physics", 85th ed., Eds. Lide, D.R., CRC Press: Boca Raton, FL, ISBN: 9781420090840

Weimlaender, M., Kennedy, E. B., Lekovic, V., Beumer, J., Moy, P. K. and Lewis, S., (1992), "Histomorphometry of bone apposition around three types of endosseous dental implants", *Inter. J. Oral Maxillofac Implants*, Vol. 7, pp. 491–496.

Weng, J., Liu, Q, Wolke, J. G. C., Zhang, X. and de Groot, K., (1997), "Formation and characteristics of the apatite layer on plasma-sprayed hydroxyapatite coatings in simulated body fluid", *Biomaterials*, Vol. 18, pp. 1027–1035.

Wu, Z-J., He, L-P. and Chen, Z-Z., (2006), "Fabrication and characterization of hydroxyapatite/Al_2O_3 biocomposite coating on titanium", *Trans. Nonferrous Met. Soc., China*, Vol. 16, pp. 259–266.

Xu, Y., Wang, D., Yang, L. and Tang, H., (2001), "Hydrothermal conversion of coral into hydroxyapatite", *Mater. Charact.*, Vol. 47 (2), pp. 83–87.

Yamada, K., Imamura, K., Itoh, H., Iwata, H. and Maruno, S., (2001), "Bone bonding behavior of the hydroxyapatite containing glass-titanium composite prepared by cullet method", *Biomaterials*, Vol. 22, pp. 2207–2214.

Zhang, C., (2000), *"Mechanical Integrity of Plasma-sprayed Hydroxyapatite Coatings on Ti-6Al-4V Implants"*, Ph.D., Thesis, Department of Mechanical Engineering, Hong Kong University of Science and Technology: Hong Kong.

Zheng, X. B., Shi, J. M. and Liu, X. Y., (2001), "Developments of plasma-sprayed biomedical coatings", *J. Ceram. Process Res.*, Vol. 2 (4), pp. 174–179.

Zheng, X., Huang, M. and Ding, C., (2000), "Bonding strength of plasma-sprayed hydroxyapatite/Ti composite coatings", *Biomaterials*, Vol. 21, pp. 841–849.

10 Characterisation and Optimisation of TiO$_2$/CuO Nanocomposite for Effective Dye Degradation from Water under Simulated Solar Irradiation

Rohini Singh
Sitarambhai Naranji Patel Institute of
Technology & Research Centre

Suman Dutta
Indian Institute of Technology (ISM)

CONTENTS

10.1 Introduction 196
10.2 Experimental 196
    10.2.1 Preparation of TiO$_2$ Nanoparticles.. 196
    10.2.2 Synthesis of TiO$_2$/CuO Nanocomposite 196
    10.2.3 Photocatalytic Activity Measurements 197
    10.2.4 Experimental Design 197
10.3 Result and Discussions 197
    10.3.1 Characteristics of TiO$_2$/CuO Nanocomposites 197
    10.3.2 Optimization Study.. 200
    10.3.3 Hydrogen Generation Capacity... 204
10.4 Conclusion 204
References 204

10.1 INTRODUCTION

Semiconductor photocatalyst has been studied extensively in solving environmental issues such as wastewater purification and solar hydrogen generation [1–3]. TiO_2 has become the most widely used photocatalyst majorly because of its high photocatalytic response and chemical stability [4,5]. However, the photocatalytic activity of TiO_2 is limited to UV radiations that constitute only 3%–5% of the solar spectrum. In addition, fast electron-hole pair recombination hinders the photoactivity of TiO_2. Therefore, effective modifications such as noble metal loading, metal cation and non-metal doping, addition of electron donors and synthesis of semiconductors composite are required [6–8]. Semiconductor composites can be utilised as an influential strategy for enhancing the overall photo response of TiO_2 [9]. CuO is a p-type semiconductor [10]. It has been identified that CuO exists in numerous forms on TiO_2 [11]. Black-coloured cupric oxide exists in the mineral phase as tenorite, and TiO_2/CuO nanoparticles exhibits enhanced photoactivity [12]. TiO_2/CuO nanocomposites offer an effective, simple and economically feasible approach to induce charge separation in TiO_2 [13]. Synergistic effect has been studied extensively by the researchers across the globe. Scuderi et al., 2017 [14] reported the enhanced photocatalytic dye degradation capability of CuO nanowires. Luna et al., 2016 [15] had successfully utilised CuO–TiO_2 composites for the photo-assisted degradation of gallic acid under UV/visible spectrum. Heterogeneous photocatalysis has been considered as an effective advanced oxidation process (AOP) that involves effective organic pollutant degradation [16].

In the present work, we attempted to synthesise, characterise and optimise TiO_2/CuO nanocomposite photocatalyst with enhanced visible spectrum activity for the effective methylene blue (MB) decomposition. Moreover, the prepared catalyst has been utilised for the production of hydrogen via photocatalytic water splitting. Therefore, the major objective of the present research is to synthesise efficient photocatalyst with dual applicability i.e. dye degradation and hydrogen generation.

10.2 EXPERIMENTAL

10.2.1 PREPARATION OF TiO_2 NANOPARTICLES

In the previous work, two different solutions of precursor, sodium hydroxide and saturated oxalic acid, were prepared and added together. Sol-gel method was used to prepare TiO_2 photocatalysts using tetrabutyl orthotitanate as a precursor [17].

10.2.2 SYNTHESIS OF TiO_2/CuO NANOCOMPOSITE

CuO was purchased from Sigma Aldrich. MB was obtained from LOBAL Chemie. The preparation of TiO_2/CuO nanocomposite photocatalyst was carried out in egoma high-energy planetary ball mill 0.3 g of TiO_2 nanoparticles with 30 g of 10 mm stainless steel (SS) balls were mixed in the 250 ml SS jar in the ratio of 1:10 and varied amount of CuO nanoparticles (0.06, 0.18 and 0.3 g) and distilled water was added. After 5 h milling at 750 rpm, the wet powder was washed thoroughly with distilled water, filtered and finally dried in the hot air oven at 100°C for 1–2 h. The prepared

TiO₂/CuO Nanocomposite

nanocomposite was then crushed into powder in mortar and then calcined at 600°C for 2 h in the muffle furnace in the presence of air.

10.2.3 PHOTOCATALYTIC ACTIVITY MEASUREMENTS

TiO_2/CuO powder of 0.5 g with varied CuO content i.e. 0.06–0.30 g was added into a 250 ml of 50 mg/l dye contaminated solution at different pH ranging from 4 to 9 based on the design of experiments via response surface method. The gaseous products were confirmed via gas chromatography. The suspensions were deaerated for 15–20 min via boiling the suspension prior to the ethanol addition to prevent the uptake of photogenerated electrons by dissolved oxygen.

10.2.4 EXPERIMENTAL DESIGN

The amount of doped CuO and pH of the dye solution was optimised via Minitab 18. The response surface methodology (RSM) was used to investigate factors in order to build models and to achieve maximum photocatalytic dye degradation [18]. The analysis of variance (ANOVA) was applied as a statistical tool to exhibit the significance and accuracy of the model. Moreover, 4D contour plot and surface plot were utilised to evaluate the optimum conditions.

10.3 RESULT AND DISCUSSIONS

10.3.1 CHARACTERISTICS OF TiO₂/CuO NANOCOMPOSITES

Optical properties of the prepared TiO_2/CuO nanocomposite photocatalysts were confirmed by the UV-visible absorption spectroscopy. Figure 10.1 shows the absorption spectra of TiO_2 derived by sol-gel method calcined at 600°C i.e. T-600 and the prepared TiO_2/CuO composite photocatalysts with varied CuO content of 0.06 and 0.3 g designated as TC (0.06) and TC (0.3), respectively. According to the absorption spectra, the prepared nanocomposites exhibit high visible light response with increased CuO content. The diffraction patterns were recorded for the confirmation of crystallite size and phases present in TiO_2/CuO composite within the scanning range (2θ) of 20°–80°. Figure 10.2 shows the X-ray diffraction (XRD) spectra of the prepared TiO_2/CuO photocatalyst with varied amount of doped CuO. Pattern for TC (0.3) and TC (0.06) reveals the presence of tenorite [JCPDS card No. # 01-1117], anatase [JCPDS card No. # 01-0562] and rutile [JCPDS card No. # 01-1292] phases in the prepared nanocomposites. Field Emission Scanning Electron Microscopy (FE-SEM) image was collected to confirm the morphology of the particles, and with the same instrument, elemental composition was investigated by energy-dispersive spectrometry (EDS) analysis. Figure 10.3 clearly describes the morphology and particle size of synthesised TiO_2/CuO nanocomposite. Our previous work illustrates the FE-SEM images of prepared TiO_2 photocatalysts [17]. Prepared catalyst exhibits almost spherical morphology with the particle size distribution in nano range. Energy-dispersive X-ray spectrum shown in Figure 10.4 was utilised to analyse the composition of prepared TiO_2/CuO sample. The carbon content may be referred

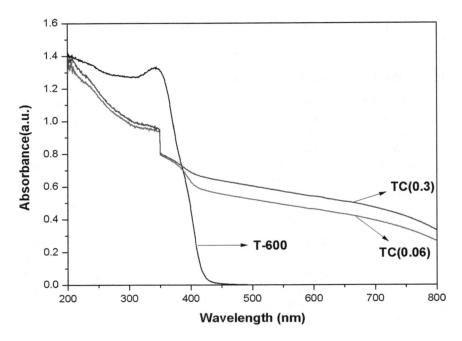

FIGURE 10.1 Absorption spectra of pure TiO_2 and composite TiO_2/CuO photocatalysts.

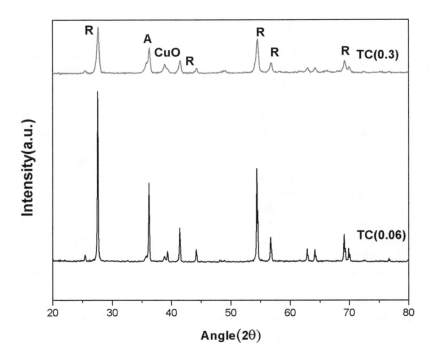

FIGURE 10.2 XRD spectra prepared TiO_2/CuO samples.

TiO₂/CuO Nanocomposite

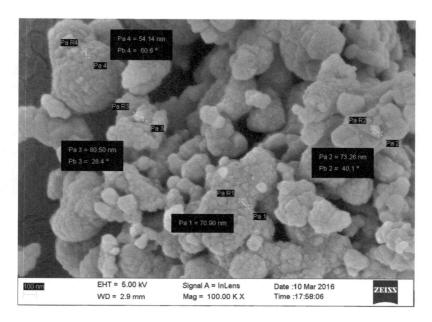

FIGURE 10.3 FE-SEM image of prepared TiO₂/CuO nanocomposite.

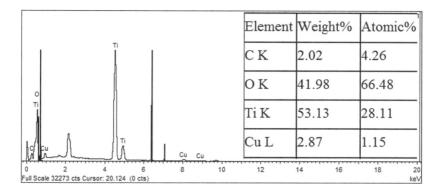

Element	Weight%	Atomic%
C K	2.02	4.26
O K	41.98	66.48
Ti K	53.13	28.11
Cu L	2.87	1.15

FIGURE 10.4 Energy-dispersive X-ray spectrum of as-prepared CuO/TiO₂ nanocomposite.

to the carbon tape used during the preparation of FE-SEM sample. Figure 10.5 a and b illustrates transmission electron microscopy (TEM) image and Selected Area Electron Diffraction (SAED) pattern of TiO₂/CuO nanoparticles by mechanical ball milling method. Based on Figure 10.5a, TiO₂/CuO composites exhibit almost spherical-shaped morphology with uniformly distributed CuO nanoparticles, shown as darker particles. Therefore, observations obtained from FE-SEM and TEM images reveal that doping has not influenced morphology of the prepared TiO₂/CuO nanoparticles. Figure 10.5b reports the presence of tenorite crystalline phase of CuO along with the rutile and anatase phases of TiO₂ in the prepared nanocomposite.

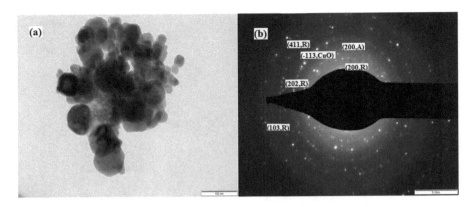

FIGURE 10.5 TEM image (a) and SAED pattern (b) of prepared TiO$_2$/CuO nanocomposite.

10.3.2 OPTIMIZATION STUDY

In the present work, the involved factors such as CuO content in the preparation of TiO$_2$/CuO nanocomposite and pH of dye solution during photocatalytic degradation were investigated, and the respective response in terms of % degradation of MB was carefully studied. As mentioned in the previous sections, 0.5 g TiO$_2$/CuO powder with varied CuO content i.e. 0.06–0.30 g was added into a 250 ml of 50 mg/l dye solution and maintained at different pH in the range of 4–9 based on the design of experiments via response surface method. Table 10.1. lists the experimental data at three levels of two independent variables, the observed response as % decomposition of MB. The degradation percentage of MB illustrated in Table 10.1 was evaluated by using the following equation.

$$\% \text{ Degradation} = \frac{C_0 - C_t}{C_0} \times 100\% \quad (10.1)$$

This clearly demonstrates that the percentage degradation of MB strongly influenced by the selected factors. Fitted second-order model shown in Equation 10.2 was formed to investigate the relation between the influencing factors and the response.

$$D_{MB} = 56.5 - 2.11*p + 270.3*a + 0.660*p^2 - 247.3*a^2 - 24.92*p*a \quad (10.2)$$

Where, D_{MB} is percentage degradation of MB, p is the dye solution pH value and a is the amount of CuO (g) utilised in the preparation of TiO$_2$/CuO nanocomposites. (p, a) indicates the variance of the responses with each single factor, whereas their quadratic term is indicated by (p^2, a^2). Moreover, ($p*a$) indicates the interaction. According to Table 10.1, experimental results were close to the predicted results. Figure 10.6 represents experimental and predicted plot of % MB degradation. ANOVA was presented in Table 10.2. Figure 10.7 a and b represents the interaction and main effect plot for the selected response i.e. % degradation of MB. The response surface illustrating the influence of factors on the response is shown in

TiO₂/CuO Nanocomposite

TABLE 10.1
Experimental Data of Two Independent Variables at Three Levels and the Observed Response in Terms of % Degradation of Methylene Blue (MB)

Exp. No.	Experimental Factors		% Degradation of Methylene Blue (MB)	
	pH of Dye Solution	Amount of Doped CuO (g)	Experimental	Predicted
1	5	0.060000	73.06	70.2903
2	9	0.060000	95.99	92.8547
3	5	0.300000	80.34	83.8878
4	9	0.300000	79.35	82.5322
5	4.17157	0.180000	81.56	81.0953
6	9.82843	0.180000	96.04	96.0922
7	7	0.0102944	70.77	75.0308
8	7	0.349706	82.02	77.4081
9	7	0.180000	83.33	83.3120
10	7	0.180000	84.21	83.3120
11	7	0.180000	82.96	83.3120
12	7	0.180000	83.06	83.3120
13	7	0.180000	83	83.3120

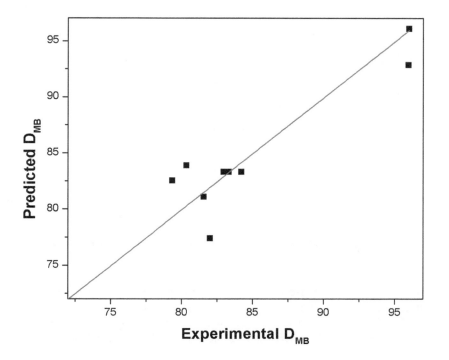

FIGURE 10.6 Experimental and predicted plot of % MB degradation.

TABLE 10.2
ANOVA of the Quadratic Model for Value of D_{MB}

Source	DF[a]	SS[b]	MS[c]	F-Value	p-Value
Model	5	529.806	105.961	9.10	0.006
Linear	2	230.272	115.136	9.89	0.009
pH of aqueous dye solution	1	224.909	224.909	19.31	0.003
Amount of CuO	1	5.363	5.363	0.46	0.519
Square	2	156.493	78.246	6.72	0.024
pH of aqueous dye solution*pH of aqueous dye solution	1	48.516	48.516	4.17	0.081
Amount of CuO*Amount of CuO	1	88.245	88.245	7.58	0.028
2-Way interaction	1	143.042	143.042	12.28	0.010
pH of aqueous dye solution*Amount of CuO	1	143.042	143.042	12.28	0.010
Error	7	81.520	11.646		
Lack-of-fit	3	80.429	26.810	98.25	0.000
Pure error	4	1.091	0.273		
Total	12	611.326			

[a] Degree of freedom
[b] Sum of squares
[c] Means square

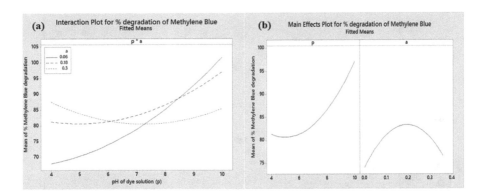

FIGURE 10.7 Interaction (a) and main effect plot (b) for the % degradation of MB.

Figure 10.8. The 3D response surface curvature illustrates the interactions between the independent variables and confirms that higher % degradation of MB can be achieved at higher pH of dye solution due to enhanced opposite charge attraction between cationic dye (MB) and the negatively charged catalyst (TiO$_2$/CuO). However, higher CuO content had the detrimental effect on the response reportedly

TiO$_2$/CuO Nanocomposite

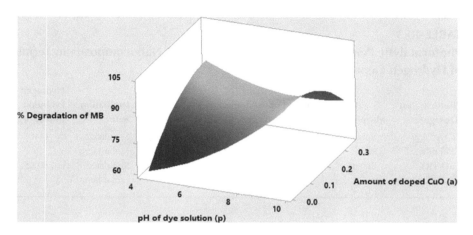

FIGURE 10.8 Response surface plots of % MB degradation.

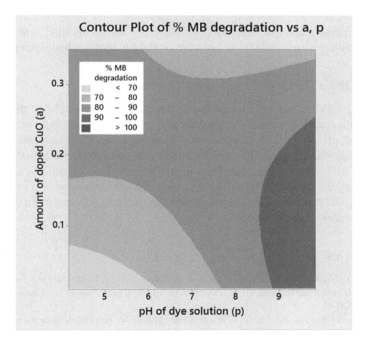

FIGURE 10.9 Response 4D contour plot of % degradation of MB.

due to the blockage of reaction sites on the catalyst surface. Figure 10.9 shows the 4D response contour plot. According to the contour plot, to obtain higher MB degradation %, the pH of dye solution should be close to nine while the amount of CuO should be maintained in the range of 0.1–0.2 g.

TABLE 10.3
Photocatalytic Performance of TiO$_2$ and TiO$_2$/CuO Nanocomposite in Terms of Hydrogen Generation

Photocatalyst (Dosage)	Morphology	Method of Preparation	Method of Modification	Sacrificial Agent	Irradiation Source	Hydrogen Evolution (mmol/h/g)
TiO$_2$ (0.5 g/ 250 ml)	Spherical	Sol-gel	Calcination (600°C)	Ethanol	Simulated solar lamp	1440
CuO/TiO$_2$ (0.5 g/250 ml)	Spherical	Sol-gel	Mechanical ball milling	Ethanol	Simulated solar lamp	2802.5

10.3.3 HYDROGEN GENERATION CAPACITY

According to the result obtained from the photocatalytic hydrogen production experiment, the synthesised photocatalysts exhibits photoactivity. Table 10.3 lists the photocatalytic performance of bare TiO$_2$ prepared by sol-gel method and TiO$_2$/CuO nanocomposite with maximum CuO content i.e. 0.3 g in terms of hydrogen generation.

10.4 CONCLUSION

In this study, TiO$_2$/CuO nanocomposite derived from mechanical ball milling process exhibits almost spherical morphology and enhanced visible light activity. CuO content and pH of the dye solution were optimised via RSM, and mathematical model was developed on the basis of statistical analysis for the prediction of % degradation of MB. CuO content and pH of the dye solution play a highly significant role in improving the photoactivity of the synthesised nanocomposites. On implementation of optimised parameters, maximum dye degradation efficiency of the prepared nanocomposites can be achieved.

REFERENCES

1. Wen, J., Li, X., Liu, W., Fang, Y., Xie, J., & Xu, Y. (2015). Photocatalysis fundamentals and surface modification of TiO$_2$ nanomaterials. *Chinese Journal of Catalysis*, 36(12), 2049–2070.
2. Chong, M.N., Jin, B., Chow, C.W., & Saint, C. (2010). Recent developments in photocatalytic water treatment technology: A review. *Water Research*, 44(10), 2997–3027.
3. Augugliaro, V., Litter, M., Palmisano, L., & Soria, J. (2006). The combination of heterogeneous photocatalysis with chemical and physical operations: A tool for improving the photoprocess performance. *Journal of Photochemistry and Photobiology C: Photochemistry Reviews*, 7(4), 127–144.
4. Gupta, S.M., & Tripathi, M. (2011). A review of TiO$_2$ nanoparticles. *Chinese Science Bulletin*, 56(16), 1639.
5. Lee, S.Y., & Park, S.J. (2013). TiO$_2$ photocatalyst for water treatment applications. *Journal of Industrial and Engineering Chemistry*, 19(6), 1761–1769.

TiO$_2$/CuO Nanocomposite

6. Anpo, M., & Takeuchi, M. (2003). The design and development of highly reactive titanium oxide photocatalysts operating under visible light irradiation. *Journal of Catalysis*, 216(1), 505–516.

7. Aysin, B., Ozturk, A., & Park, J. (2013). Silver-loaded TiO$_2$ powders prepared through mechanical ball milling. *Ceramics International*, 39(6), 7119–7126.

8. Shifu, C., Lei, C., Shen, G., & Gengyu, C. (2005). The preparation of coupled WO$_3$/TiO$_2$ photocatalyst by ball milling. *Powder Technology*, 160(3), 198–202.

9. Ni, M., Leung, M.K., Leung, D.Y., & Sumathy, K. (2007). A review and recent developments in photocatalytic water-splitting using TiO$_2$ for hydrogen production. *Renewable and Sustainable Energy Reviews*, 11(3), 401–425.

10. Yu, Z., Meng, J., Li, Y., & Li, Y. (2013). Efficient photocatalytic hydrogen production from water over a CuO and carbon fiber comodified TiO$_2$ nanocomposite photocatalyst. *International Journal of Hydrogen Energy*, 38(36), 16649–16655.

11. Chen, W.T., Jovic, V., Sun-Waterhouse, D., Idriss, H., & Waterhouse, G.I. (2013). The role of CuO in promoting photocatalytic hydrogen production over TiO$_2$. *International Journal of Hydrogen Energy*, 38(35), 15036–15048.

12. Lee, S.S., Bai, H., Liu, Z., & Sun, D.D. (2013). Optimization and an insightful properties—Activity study of electrospun TiO$_2$/CuO composite nanofibers for efficient photocatalytic H$_2$ generation. *Applied Catalysis B: Environmental*, 140, 68–81.

13. Kum, J.M., Yoo, S.H., Ali, G., & Cho, S.O. (2013). Photocatalytic hydrogen production over CuO and TiO$_2$ nanoparticles mixture. *International Journal of Hydrogen Energy*, 38(31), 13541–13546.

14. Scuderi, V., Amiard, G., Sanz, R., Boninelli, S., Impellizzeri, G., & Privitera, V. (2017). TiO$_2$ coated CuO nanowire array: Ultrathin p–n heterojunction to modulate cationic/anionic dye photo-degradation in water. *Applied Surface Science*, 416, 885–890.

15. Luna, A.L., Valenzuela, M.A., Colbeau-Justin, C., Vázquez, P., Rodriguez, J.L., Avendano, J.R., … José, M. (2016). Photocatalytic degradation of gallic acid over CuO–TiO$_2$ composites under UV/Vis LEDs irradiation. *Applied Catalysis A: General*, 521, 140–148.

16. Khaki, M.R.D., Shafeeyan, M.S., Raman, A.A.A., & Daud, W.M.A.W. (2017). Application of doped photocatalysts for organic pollutant degradation-A review. *Journal of Environmental Management*, 198, 78–94.

17. Singh, R., & Dutta, S. (2017). Synthesis and characterization of solar photoactive TiO$_2$ nanoparticles with enhanced structural and optical properties. *Advanced Powder Technology*, 29, 211–219.

18. Bezerra, M.A., Santelli, R.E., Oliveira, E.P., Villar, L.S., & Escaleira, L.A. (2008). Response surface methodology (RSM) as a tool for optimization in analytical chemistry. *Talanta*, 76(5), 965–977.

11 Dual Applicability of Hexagonal Pyramid-Shaped Nitrogen-Doped ZnO Composites As an Efficient Photocatalyst

Rohini Singh
Sitarambhai Naranji Patel Institute of
Technology & Research Centre

Suman Dutta
Indian Institute of Technology (ISM)

CONTENTS

11.1 Introduction 208
11.2 Experimental Procedure...208
    11.2.1 Chemicals 208
    11.2.2 Preparation of p-ZnO Photocatalyst 209
    11.2.3 Characterization of Photocatalyst...209
    11.2.4 Photoactivity Measurements..210
        11.2.4.1 Dye Decomposition..210
        11.2.4.2 Hydrogen Production 210
11.3 Result and Discussions 210
    11.3.1 Characteristics of Photocatalyst 210
        11.3.1.1 UV-Visible Absorption Spectroscopy 210
        11.3.1.2 X-ray Diffraction 210
        11.3.1.3 Field Emission Scanning Electron Microscopy
                (FE-SEM) 211
        11.3.1.4 Energy-Dispersive X-ray Analysis..................................212
        11.3.1.5 Particle Size Distribution (PSD) Analysis 212
        11.3.1.6 X-Ray Photoluminescence (PL) Spectroscopy 213
    11.3.2 Photoactivity of N/ZnO 213
        11.3.2.1 Dye Decomposition..213
        11.3.2.2 Hydrogen Generation 215
11.4 Conclusion 216
References..216

11.1 INTRODUCTION

Effluents of textile dye industries are nowadays considered as one of the major hazardous organic compound sources. Dyeing procedure adopted in the textile industries produces the significant amount of dye contaminated wastewater worldwide. Heterogeneous photocatalysis using solar energy and ZnO has the capability for hazardous pollutants abatement and solar hydrogen generation [1].

ZnO possesses excellent features such as enhanced chemical stability and mobility of electrons along with unique piezoelectric property. It is used in the fabrication of gas and chemical sensors, biosensors and transducers [2–6]. Various techniques such as sol-gel method, hydrothermal method along with vapour deposition via chemical and physical methodology had been developed for the synthesis of ZnO [7–10]. ZnO is a promising photocatalyst that is ultraviolet radiation active that contributes approximately 3%–5% of the solar spectrum only. Thus, to enhance the visible radiation activity of ZnO, different methods are adopted such as –ion doping, non-metal doping, co-doping and dye sensitization [11–13]. Nitrogen is considered as a promising dopant to synthesise p-type ZnO [14].

In the photon-assisted catalytic reactions, a major role has been played by the electronic structure of prepared ZnO consisting of conduction and valence band. The excitation of electrons gets executed on receiving the photons with the energy greater or equal than E_g i.e. band gap.

The overall photocatalytic dye degradation process can be illustrated in the form of the following equation [15,16]:

$$ZnO + h\nu \rightarrow e^- + h^+ + ZnO \qquad (11.1)$$

Although the photo-assisted catalytic dye decomposition is a clean technology and utilises renewable resources such as water and sunlight, visible light-responsive photocatalyst needs to be synthesised. It is essential to use visible light efficiently to realise solar hydrogen production on a large scale since the major portion of the solar radiation consists of visible light.

In this chapter, visible light active p-type N/ZnO (nitrogen-doped) photocatalyst was synthesised via zinc nitrate hexahydrate decomposition. Though some researchers worked on same process but in the present research, characterization and photocatalytic dye degradation efficiency analysis of the prepared ZnO catalyst were carried out in more detail. In addition to that, volume of hydrogen produced in 2 h was estimated utilising prepared ZnO as photocatalyst. Transition metal oxides act as efficient photocatalysts [17–19].

11.2 EXPERIMENTAL PROCEDURE

11.2.1 CHEMICALS

Commercial ZnO, Zinc nitrate hexahydrate and ethanol were purchased from Merck. Rhodamine 6G and methylene blue were procured from LOBAL Chemie. Eosin Y was purchased from Sigma Aldrich. Table 11.1 shows the list of structures and properties of the dyes utilised in the present work.

Efficient Photocatalyst

TABLE 11.1

List of Structure and Properties of the Dyes Utilised in the Present Work

Dyes	Chemical Formula	Nature	λ_{max} (nm)	Molecular Weight (g/mol)	Molecular Structure
Methylene blue	$C_{16}H_{18}ClN_3S$	Cationic	664	319.85	
Rhodamine 6G	$C_{28}H_{31}N_2O_3Cl$	Cationic	522–527	479.02	
Eosin Y	$C_{20}H_8Br_4O_5$	Anionic	524	647.89	

11.2.2 PREPARATION OF P-ZNO PHOTOCATALYST

The p-type nitrogen-doped ZnO catalyst was synthesised by the thermal degradation of zinc nitrate hexahydrate. The sample was calcined at 10°C/min–12°C/min in the presence of air using muffle furnace till 400°C and treated for 30 min to ensure proper nitrate decomposition. Further, the prepared sample was ball milled for 1/2 h at 500 rpm for desired size reduction and increased surface area.

11.2.3 CHARACTERIZATION OF PHOTOCATALYST

Field emission scanning electron microscopy was utilised to confirm the particles structure and morphology, whereas the composition of the synthesised photocatalyst was confirmed via energy-dispersive spectrometry. The optical property of the N/ZnO was observed with ultraviolet (UV)-visible spectrophotometer. The crystal size and phase of the synthesised ZnO were analysed by X-ray diffraction technique. Particle size distribution data were recorded in Microtrac S3500 particle size analyzer. Photoluminescence (PL) spectroscopy was carried out in F-2500 Fluorescence Spectrophotometer to explain the concept of electron–hole pairs recombination and separation involved in the overall mechanism of photocatalysis.

11.2.4 PHOTOACTIVITY MEASUREMENTS

11.2.4.1 Dye Decomposition

The catalytic response of synthesised ZnO sample was estimated via monitoring the photon degradation of Rhodamine 6G (Rh 6G), methylene blue (MB) and Eosin Y in aqueous solution. Solar simulated radiations of 300 W were utilised as the light source and UV-visible spectrophotometer was utilised for the purpose of photocatalytic activity evaluation. About 1 g p-type ZnO powder was added into a 250 ml of 50 mg/L dye solution maintained at the pH of 10 for MB and Rh 6G, whereas 4 in case of Eosin Y. The solution was continuously magnetic stirred at the rpm of 500 in dark for the duration of 60 min for the absorption of dye solution on the prepared ZnO powder to attain absorption equilibrium. In the present paper, we will consider N/ZnO for the degradation of MB, Rh 6G and Eosin Y as N/Zn_MB, N/Zn_Rh and N/Zn_EY, respectively.

11.2.4.2 Hydrogen Production

Magnetic stirrer was used in the experiment for continuous stirring. A round bottom flask was covered with rubber septum. About 0.5 g of synthesised catalyst was mixed into 250 ml distilled water with ethanol as a sacrificial agent. Gas sample of 1 ml was taken after every 2 h of continuous irradiation via an airtight syringe from the rubber septum for chromatographic measurements.

11.3 RESULT AND DISCUSSIONS

11.3.1 CHARACTERISTICS OF PHOTOCATALYST

11.3.1.1 UV-Visible Absorption Spectroscopy

Figure 11.1 shows the absorption graph of the prepared p-type N/ZnO photocatalyst and commercial zinc oxide i.e. ZnO(C). The calculated band gap of the synthesised and commercial ZnO photocatalyst evaluated by Kubelka–Munk function was found to be 3.1 and 3.5 eV, respectively.

11.3.1.2 X-ray Diffraction

The X-ray diffraction (XRD) peaks were obtained to confirm the crystallite phase and size within the scanning range (2θ) of $10°–80°$. Figure 11.2 shows the XRD spectra of the prepared p-type ZnO photocatalyst. Diffraction analysis confirms the presence of wurtzite ZnO mineral phase (JCPDS 36-1451) with the average crystallite size of 42.87 nm. The average crystallite size of the prepared N/ZnO was calculated by taking an average of the major peaks of the diffraction data obtained via the Scherer's equation.

$$D = \frac{K\lambda}{\beta \cos\theta} \qquad (11.2)$$

Efficient Photocatalyst

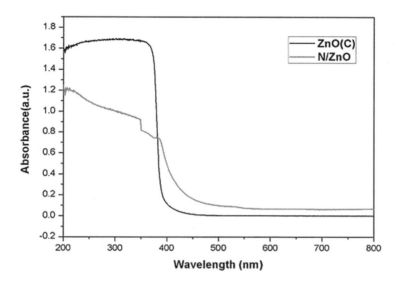

FIGURE 11.1 Absorption spectra of the prepared (N/ZnO) and commercial ZnO photocatalyst.

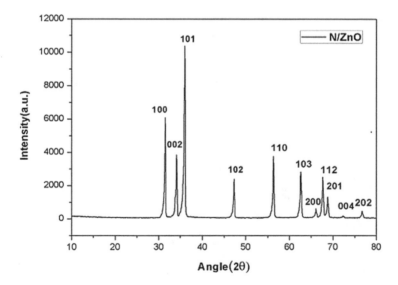

FIGURE 11.2 X-ray diffraction spectra of prepared nano-crystalline N/ZnO photocatalyst.

11.3.1.3 Field Emission Scanning Electron Microscopy (FE-SEM)

FE-SEM images were obtained for the confirmation of morphology of the particles and elemental composition was also analysed via EDAX. Figure 11.3 clearly describes the hexagonal pyramid morphology of synthesised ZnO sample.

FIGURE 11.3 FE-SEM images of prepared ZnO nanoparticles (a) and (b) at different magnification.

11.3.1.4 Energy-Dispersive X-ray Analysis

Figure 11.4 illustrates the energy-dispersive X-ray (EDX) spectrum of the 67.26 wt. % Zn, 20.53 wt. % O, 1.32 wt. % N, 7.63 wt. % Pt and 3.28 wt. % C. The peaks of Zn and O facilitated firm evidence that the prepared photocatalyst consists of only ZnO.

11.3.1.5 Particle Size Distribution (PSD) Analysis

Figure 11.5 reports the particle size distribution (PSD) of N/ZnO photocatalyst. It can be observed from the PSD chart, that particle size of the prepared sample lies in the range of 0.6–350 μm. Table 11.2 illustrates the statistical data of prepared N/ZnO photocatalyst.

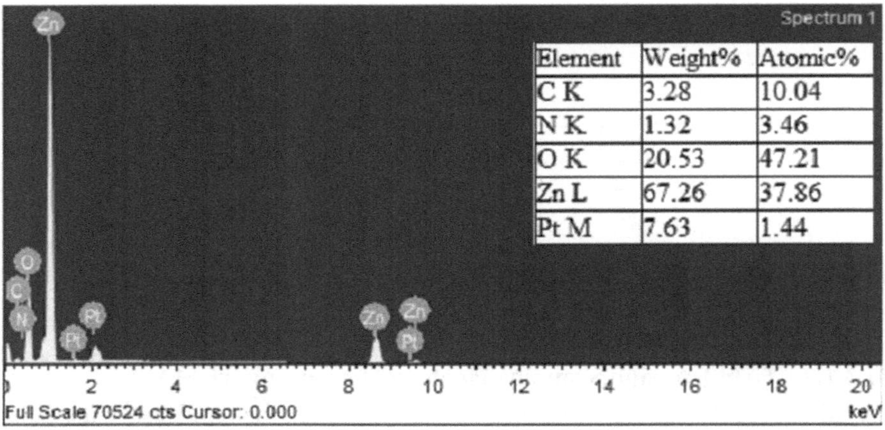

Element	Weight%	Atomic%
C K	3.28	10.04
N K	1.32	3.46
O K	20.53	47.21
Zn L	67.26	37.86
Pt M	7.63	1.44

FIGURE 11.4 Energy-dispersive X-ray spectrum of ZnO photocatalyst.

Efficient Photocatalyst

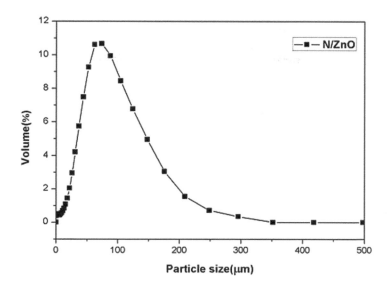

FIGURE 11.5 Particle size distribution of the prepared N/ZnO photocatalyst.

TABLE 11.2
Statistical Data of the Prepared N/ZnO

N (Number of Samples)	125 (μm)
Mean	19.40722
Mode	14.709
Median	18.111
Minimum	6.466
Maximum	42.567
Standard deviation	7.1694

11.3.1.6 X-Ray Photoluminescence (PL) Spectroscopy

As shown in Figure 11.6, samples exhibit significant PL. The PL spectroscopy intensity of N/ZnO is less than that of commercial ZnO.

11.3.2 PHOTOACTIVITY OF N/ZnO

11.3.2.1 Dye Decomposition

Rh 6G, MB and Eosin Y were used as the target dyes for the estimation of the photoactivity of the prepared N/ZnO sample under simulated solar irradiation. The efficiency of photocatalytic degradation described in Figure 11.7 was calculated by using the following equation.

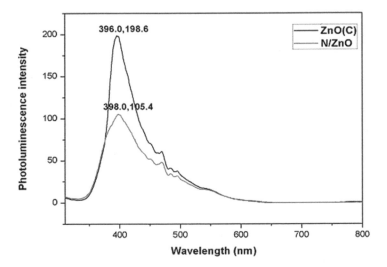

FIGURE 11.6 Photoluminescence spectra of N/ZnO and commercial ZnO photocatalyst.

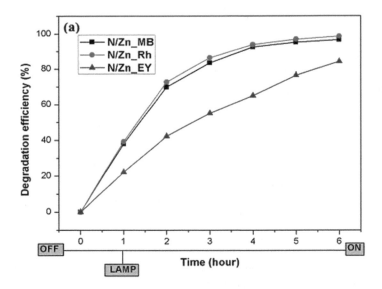

FIGURE 11.7 Photocatalytic degradation efficiency of MB, Rh 6G and Eosin Y under simulated solar irradiation over N/ZnO photocatalysts.

$$\text{Degradation Efficiency } (\%) = \frac{C_0 - C_t}{C_0} \times 100\% \qquad (11.3)$$

The kinetics of photodegradation of the utilised model dyes on the surface of the photocatalyst can be described via first-order reaction:

Efficient Photocatalyst 215

$$\ln \frac{C_0}{C_t} = kt \qquad (11.4)$$

Figure 11.8 illustrates the linearity of $\ln(C_0/C_t)$ versus irradiation time (t). Rate constants are calculated for all the analysed samples from the slope of linear fitting regression line as shown in Figure 11.8. Table 11.3 shows the rate constants for all the experimental samples.

11.3.2.2 Hydrogen Generation

Hydrogen generation (mmol/h/g) was calculated using ideal gas law as shown in Equation (11.5).

$$PV = nRT \qquad (11.5)$$

The presented hydrogen generation experiments were executed at 1 atm, room temperature (30°±2°), and the volume taken in airtight syringe was 1 ml or 0.001 l.

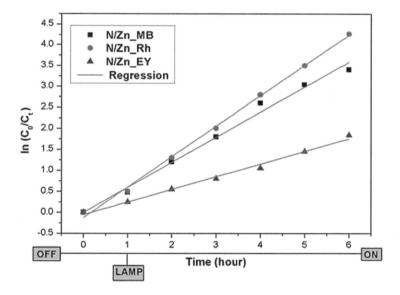

FIGURE 11.8 Kinetics of the photodegradation of MB, Rh 6G and Eosin Y in aqueous solution under simulated solar irradiation.

TABLE 11.3
Rate Constant of MB, Rh 6G and Eosin Y Photocatalytic Degradation for the Synthesised N/ZnO and Commercial ZnO Samples

Sample	N/Zn_MB	N/Zn_Rh	N/Zn_EY
k(h⁻¹)	0.5971	0.7254	0.3007

The overall hydrogen generated was then multiplied by 250 ml i.e. the empty volume of the round bottom flask available for the generated hydrogen to fill, assuming the uniform distribution of the gas. Photo-assisted catalytic hydrogen production at the pH value of 4 over 0.5 g of N/ZnO is 1827.5 mmol/h/g.

11.4 CONCLUSION

- N/ZnO derived from nitrate degradation method exhibits hexagonal pyramid morphology with the particle size in the range of 0.6–350 μm.
- The evaluated band gap value of the synthesised N/ZnO photo-assisted catalyst estimated by Kubelka–Munk function was calculated to be 3.1eV and possesses increased suppression of electron–hole pair recombination as compared to commercial ZnO.
- ZnO catalyst possesses dual applicability as an efficient photo-responsive catalyst or photocatalyst. The dye degradation efficiency of hexagonal pyramid-shaped N/ZnO photocatalyst is in the order of N/Zn_Rh> N/Zn_MB> N/Zn_EY.
- Prepared N/ZnO can be effectively utilised for photocatalytic hydrogen generation.

REFERENCES

1. R. Saleh, N.F. Djaja, Transition-metal-doped ZnO nanoparticles: Synthesis, characterization and photocatalytic activity under UV light, *Spectrochimica Acta Part A: Molecular and Biomolecular Spectroscopy* 130 (2014) 581–590.
2. S.Y. Guo, T.J. Zhao, Z.Q. Jin, X.-M. Wan, P.G. Wang, J. Shang, S. Han, Self-assembly synthesis of precious-metal-free 3D ZnO nano/microspheres with excellent photocatalytic hydrogen production from solar water splitting, *Journal of Power Sources* 293 (2015) 17–22.
3. P.K. Samanta, A. Saha, T. Kamilya, Morphological and optical property of spherical ZnO nanoparticles, *Optik* 126 (2015) 1740–1743.
4. P.K. Samanta, S. Mishra, Solution phase synthesis of ZnO nanopencils and their optical property, *Materials Letters* 91 (2013) 338–340.
5. J. Li, R. Kykyneshi, J. Tate, A.W. Sleight, p-Type zinc oxide powders, *Solid State Sciences* 9 (2007) 613–618.
6. R. Singh, S. Dutta, The role of pH and nitrate concentration in the wet chemical growth of nano-rods shaped ZnO photocatalyst, *Nano-Structures & Nano-Objects* 18 (2019) 100250.
7. S. Liao, H. Donggen, D. Yu, Y. Su, G. Yuan, Preparation and characterization of ZnO/TiO_2, $SO_4^{2-}/ZnO/TiO_2$ photocatalyst and their Photocatalysis, *Journal of Photochemistry and Photobiology A: Chemistry* 168 (2004) 7–13.
8. S. Shit, T. Kamilya, P.K. Samanta, A novel chemical reduction method of growing ZnO nanocrystals and their optical property, *Materials Letters* 118 (2014) 123–125.
9. Y. Wang, Y. Yang, L. Xi, X. Zhang, M. Jia, H. Xu, H. Wu, A simple hydrothermal synthesis of flower-like ZnO microspheres and their improved photocatalytic activity, *Materials Letters* 180 (2016) 55–58.
10. H. Wang, J. Xie, K. Yan, M. Duan, Growth mechanism of different morphologies of ZnO crystals prepared by hydrothermal method, *Journal of Materials Science & Technology* 27(2), (2011) 153–158.

11. Y. Yang, H. Li, F. Hou, J. Hu, X. Zhang, Y. Wang, Facile synthesis of ZnO/Ag nanocomposites with enhanced photocatalytic properties under visible light, *Materials Letters* 180 (2016) 97–100.
12. J. Zhang, W. Wang, X. Liu, Ag–ZnO hybrid nanopyramids for high visible-light photocatalytic hydrogen production performance, *Materials Letters* 110 (2013) 204–207.
13. R. Singh, S. Dutta. Visible light active nanocomposites for photocatalytic applications. K. Kumar and J. Paulo Davim (Eds.) In *Composites and Advanced Materials for Industrial Applications* (2018, pp. 270–296). Pennsylvania: IGI Global.
14. C. Shifu, Z. Wei, L. Wei, Z. Sujuan, Preparation, characterization and activity evaluation of p-n junction photocatalyst p-ZnO/n-TiO$_2$, *Applied Surface Science* 255 (2008) 2478–2484.
15. S. Akir, A. Barras, Y. Coffinier, M. Bououdina, R. Boukherroub, A.D. Omrani, Eco-friendly synthesis of ZnO nanoparticles with different morphologies and their visible light photocatalytic performance for the degradation of Rhodamine B, *Ceramics International* 42 (2016) 10259–10265.
16. I.M. Pereira Silva, G. Byzynski, C. Ribeiro, E. Longo, Different dye degradation mechanisms for ZnO and ZnO doped with N (ZnO:N), *Journal of Molecular Catalysis A: Chemical* 417 (2016) 89–100.
17. R. Singh, S. Dutta, A review on H$_2$ production through photocatalytic reactions using TiO$_2$/TiO$_2$-assisted catalysts, *Fuel* 220 (2018) 607–620.
18. R. Singh, S. Dutta, Synthesis and characterization of solar photoactive TiO$_2$ nanoparticles with enhanced structural and optical properties, *Advanced Powder Technology* 29 (2018) 211–219.
19. R. Singh, P. Kumari, P. D. Chavan, S. Datta, S. Dutta, Synthesis of solvothermal derived TiO$_2$ nanocrystals supported on ground nano egg shell waste and its utilization for the photocatalytic dye degradation, *Optical Materials* 73 (2017) 377–383.

Index

AC-AC converters 101
AC-DC converters 101
ACP *see* amorphous calcium phosphate (ACP)
additive manufacturing 121
advanced oxidation process (AOP) 194
aero-engines 110, 111
aircraft 109
    aero-engines 110, 111
    demands/improvements 110–111
alginate 23–24, 31
aliphatic polyesters 145
alkylene oxalates 28
alloys 111
    Al-7Si alloy 126, 127
    anode 88
    aluminum (*see* aluminum alloy)
    A356 aluminum 124
    Co-Cr 165
    lithium-metal 89
    metallic glass 46
    morphology 123
    Sn-15Pb 126, 127
Al-7Si alloy 126, 127
aluminum alloy 71
    morphology 128
    rheology (*see* rheology)
    RSF 128–129
    shear rate 128–129
    solid fraction 126
    temperature 126–128
aluminum-based self-lubricating composite
        material (Al-SLMMC) 66
    advantages/disadvantages 67–68
    coefficient of friction, composites 77, 78
    composition of reinforcement 77
    development stages 67–68
    dry sliding behavior 80
    fabrication methods 73–74
    GCI 69
    historical background 67–68
    hybrid MMC 80
    in-organic material 78
    mechanical properties 77
    mechanical *vs.* tribological aspect 76, 78,
        80, 81
    processing techniques 74
    reinforcement 74–76
    SEM images 79
    surface modification 81
    tensile modulus 78
    weight percentage 79

amorphous calcium phosphate (ACP)
    149, 170
amorphous semiconductors
    applications 59–60
    atomic orientations 42–43
    band structures
        CFO model 48, 49
        Davis–Mott model 48, 50
        MDS model 48–51
    components 38
    characteristics of 41–42
    classifications of 44, 47
    vs. crystalline solids 40–42
    crystallite nucleation 44
    CVD 53
    Debye–Scherrer relation 44
    diffraction 44
    experiments (*see* electrical characterization,
        optical characterization)
    flash evaporation technique 52
    glow discharge decomposition technique 53
    mechanical milling 53–54
    micro-scale/nano-scale level 43
    vs. non-crystalline materials 43–44
    PLD 53
    quenching technique 51
    RDF 43
    research problems 61
    semiconducting properties 44
    site-map 45
    sputtering 52–53
    SRO 38–39
    thermal evaporation 51–52
analysis of variance (ANOVA) 195
anode materials, LiBs 88
anodes 87–89
Aoki, H. 172
AOP *see* advanced oxidation process (AOP)
Armand, M. 87, 95
A356 aluminum alloy 124
atomic orientations 42–43
Avendano, E. 10

Balamurugan, A. 173
band structures
    CFO model 48, 49
    Davis–Mott model 48, 50
    MDS model 48–51
battery energy storages (BES) 100
BCT *see* bodycentred tetragonal (BCT)
Berzelius 4

219

Index

220

binder design
    adherence/mechanical stability 91–92
    electrical/ionic conductivity 90–91
    electrochemical stability 90
    electrolyte interaction 92
    physical/chemical constraints 90
    SEI formation 92
    Si anodes 92
Bingham fluid 125
bioactive fixation 167
biodegradable hydrogels 136
biological fixation 166–167
biomaterials
    CaP (*see* calcium phosphate nanoparticles
        (CaP))
    ceramic 135
    definition 163
    eye 163
    history 164
    natural (*see* natural polymeric biomaterials)
    orthopaedic/dental implant 164–179 (*see also*
        orthopaedic/dental implant materials)
    polymer (*see* polymeric biomaterials)
    polymer-based 134
    teeth 163
    tissue engineering 136–139
Birla, R. K. 31
black-coloured cupric oxide 194
bodycentred tetragonal (BCT) 113
Bohner, M. 183
bond/intermediate coat 175–176
Buma, P. 179

calcium orthophosphates 167–168
calcium phosphate bioceramic scaffolds 135
calcium phosphate nanoparticles (CaP)
    ACP 170
    apatite 170–171
    biological *vs.* synthetic HA 179
    bond coat 175–176
    bone implant interface 176–178
    bone regeneration 148–149
    ceramics 168
    clinical dentistry 149–150
    HA-based composites 172–175
    HA-coated implant 177, 178
    hydroxyapatite 171
    phosphate minerals 169
    porous HA ceramics 171–172
    TCP 169
    TTCP 169–170
carbon nanotubes (CNT) 144, 175
carboxymethyl cellulose (CMC) 92, 141
Carpenter, M. K. 10
casting 74, 75
cathode
    carbon-based anodes 102

high voltage stability 91
    LFP 87
    $LiCoO_2$ 86
    LMO 87
    OLO 86
    pathways, energy storage 85
    transition metal oxides 85
cell adhesion 144, 176
ceramic biomaterials 135
ceramic electrolytes (CEs) 96
ceramics 21, 121, 134
chalcogenide glasses 39, 59
chalcogenide glassy semiconductors (CGS)
    45–46
Chang, E. 175
chemical hydrogels 140–141
chemical vapor deposition (CVD) 53
chemokine-reinforced hydrogel 141
chitin 23
chitosan 23
    composition 141
    conjugated formulations 141–142
    tissue engineering application 142–143
chitosan-based hydrogels 139
chondrocytes proliferation 135
Chou, B.-Y. 175
Chow, L. C. 182
chromogenic materials 3
CMC *see* carboxymethyl cellulose (CMC)
CNT *see* carbon nanotubes (CNT)
cobalt–chromium (Co-Cr) alloys 164
Co-based superalloys 114–115
Cohen, M. H. 43, 48
collagen 21–22
composite material 66
composite polymer electrolytes (CPE) 95
conductive polymers 7, 8
conventional press-and-sinter process 116–117
corrosion-resistant materials 66
covalent amorphous semiconductors
    CGS 45–46
    TAS 45
    tetrahedral glasses 46
crystalline semiconductors 40–42
CVD *see* chemical vapor deposition (CVD)

dangling bonds 40
Das, P. 128
Debye–Scherrer relation 44
De Groot, K. 182
degumming 144
depth of discharge (DOD) 100
Diaz, A. F. 7
dicalcium phosphate (DCP) 180
Diesbach 4, 7
Di Palma, J. A. 175
directional solidification 116

Index

dispersive AC loss 54
dissolution/reprecipitation process 178
Ducheyne, P. 179
dye decomposition 208

ECM *see* extracellular matrix (ECM)
EDS *see* energy-dispersive spectrometry (EDS)
electrical characterization
    AC conductivity 54
    DC conductivity 54
    defect state measurements 55
electric cooling energy 3
electric lighting energy 3
electric vehicles (EV) 102
electrochemical energy 84
electrochromic oxide films
    anodic coloured material 9–10
    cathodic coloured material 10–12
electrochromic (EC) phenomenon 2
electrochromic (EC) materials
    electrochromic layer 7–9
    electrolyte 6–7
    energy efficiency 2–4
    transparent conductive electrodes 5–6
electrochromism 4
electrolytes
    aqueous medium 6
    carbonate-based 90
    CE 96
    CPE 95
    definition 92
    features 93
    ionic conductivity 97–98
    ISE 96–97
    liquid *vs.* solid 86
    liquid/quasi-solid electrolytes 94
    organic liquid non-aqueous electrolytes 93–94
    PEO 93
    solid 85
    solid polymer electrolytes 94–95
    types 6
energy-dispersive spectrometry (EDS) 195
energy-dispersive X-ray analysis (EDX) 210
energy harvester
    all-solid-state batteries 102–103
    battery chemistries 101–102
    battery-type market shares 102
    definition 98
    Li-ion batteries 99–100
    power management 100–101
    sources 98
erosive wear 72
EV *see* electric vehicles (EV)
extracellular matrix (ECM) 135

fabrication 115
    directional solidification 116

investment casting 115–116
powder metallurgy
    additive manufacturing 121
    conventional press-and-sinter process
        116–117
    HIP 117, 118
    microwave sintering 119, 120
    MIM 119, 120
    SHS 117–119
    SPS 119
face-centred cubic (FCC) 112
Fazan, F. 182
Fe-Ni-based superalloys 114
fibroblast growth factor (FGF) 31
fibroin-based biomaterial 143–144
fibronectin 22–23
fibrous capsule, formation 177
field emission scanning electron microscopy
    (FE-SEM) 209, 210
flash evaporation technique 52
fluid behavior, types 125
foil-based electrochromic device 5
fossil fuels 2
Fu, L. 174

gelatin 29–30, 140
glass-forming solids 39
Gottlander, M 173
Granath 128
Granqvist, C. -G. S. 10
graphite electrode 88
gray cast iron (GCI) 69
Gu, Y. W. 174

Hamdi, M. 183
Hench, L. L. 166, 167, 172
Herschel, W. H. 4
Herschel–Bulkley model 125, 126
highest occupied molecular orbitals (HOMO)
    90, 91
high-stress wear test 73
hip prosthesis implant 166
hot isostatic pressing (HIP) 117, 118
Hulbert, S. F 175
human Mesenchymal Stem Cells (hMSCs) 30
hyaluronan 24–25
hydrogels 135, 139
hydrogen generation capacity 202
hydrogen production 208
hydroxyapatite (HA) 150, 171
    calcium phosphorus ceramics 168
    chemical reactions 182
    chitin 142
    degradation 184–185
    dissolution behaviour 180–183
    equilibrium diagram 181
    failure mechanism 179–180

hydroxyapatite (HA) (*cont.*)
 heavy metals 184
 natural *vs.* synthetic 179
 PHB 147
 physical/chemical properties 30
 osteo-conductivity efficacy 142
 solubility isotherms 180
 structure 171

Ide-Ektessabi, A. 183
Inagaki, M. 174
inorganic solid electrolytes (ISE) 93, 96–97
investment casting 115–116
ionic amorphous solids 46
ionic conductive membranes 6
ionic liquids 7
Ismail, M. H. 116
Iwasaki, N. 142

Kannan, S. 183
Kim, H. 176
Koke, J. 125
Kolomiets, B. T. 39, 46
Kurzweg, H. 176

Lampert, C. M. 10
Lansaker, P. C. 5
largescale installations 100
laser drilling 121
Lee, T. M. 173
Leftheriotis, G. 6
Lemaitre, J. 183
Li, H. 174
ligature 164
Lim, V. J. P. 174
liquid electrolytes 84
liquid/quasi-solid electrolytes 94
liquid *vs.* solid electrolytes 86
lithium-ion batteries (LiBs) 84, 85
lithium-metal alloys 89
Livingston 172
Lorenz 126
lost wax casting 115
low-emissivity windows 2–3
lowest unoccupied molecular orbitals (LUMO)
 90, 91
low-stress wear test 73
lubricant material 66
Luna, A. L. 194

macroporous 171
magnesium 89
Marguis, P. M. 182
McManus et al. 31
mechanical milling (MM) 53–54
Mehrabian, R. 128
metal-based materials 134

metal injection moulding (MIM) 119, 120
metallic amorphous solids 46
metal matrix layer (MML) 66
metal oxides 6, 8
methylene blue (MB) 194
microelectromechanical system 67
micro grids 98
microporous material 171
microwave sintering 119, 120
Modigell, M. 123, 125
morphological fixation 166
Mott, N. F. 46, 48, 49, 50
musculoskeletal system 176–177
Muthukumaran, V. 183

Nagano, M. 177, 182
nanotechnology 1
 scaffold-based manipulation 134
 therapeutic strategies 134
 tissue engineering 134
natural materials 1
natural polymeric biomaterials
 medical application 25
 polysaccharides
  alginate 23–24
  chitin/chitosan 23
  hyaluronan 24–25
 proteins
  collagen 21–22
  fibronectin 22–23
  silk 22
 tissue engineering 28
natural polymers 135, 136
Newtonian fluids 124
Ni-based superalloys 112–113
nickel-cadmium (Ni-Cd) 84
nickel-metal hydride (Ni-MH) 84
Nie, X. 176
non-dendritic morphology 124
non-Newtonian fluids 124–126
normal hydrogen electrode (NHE) 99

OHA *see* oxy-hydroxyapatite (OHA)
OLO *see* over-lithiated oxides (OLO)
one-pair semiconductors 45
optical characterization
 absorption process 56
 disorderness 55
 DOS 58
 OJL model 57
 perturbation 57
 thin-film samples 55
 transmittance 55
organic liquid non-aqueous electrolytes 93–94
orthopaedic/dental implant materials
 bioactive fixation 167
 biological fixation 166–167

Index

body system 165
cement-less procedures 166
hip prosthesis implant 166
implant loosening 168
metallic implants 164–167
morphological fixation 166
properties used 165
osteolytic response 166
over-lithiated oxides (OLO) 86
Ovshinsky, S. R. 39, 48
oxy-hydroxyapatite (OHA) 181

Paper, L. 123
partial state of charge (PSOC) 101
particle size distribution (PSD) analysis 210–211
PCL *see* polycaprolactone (PCL)
PEG *see* poly-ethylene glycol (PEG)
Peled, E. 88
photochromic materials 3
photon-assisted catalytic reactions 206
physical hydrogels 139–140
pin disc apparatus 73
polyacrylates 26, 27
polyanhydrides 30
polycaprolactone (PCL) 148
polyesters 26, 145
 PHAapplications 145–147
 PHBapplications 147–148
poly-ethylene glycol (PEG) 30
polyethylene oxide (PEO) 6, 93
polyethylene terephthalate (PET) 5
poly(glycerol sebacate) (PGS) 147
polyglycolic acid (PGA) 28
polyhydroxybutyrate (PHB) 136, 147–148
polylactic acid (PLA) 28
polylactide-co-glycolide (PLGA) 135, 149
polymer-based biomaterials 134
polymeric biomaterials
 vs. ceramic materials/metals 21
 natural 21–25 (*see also* natural polymeric
  biomaterials)
 synthetic 25–28 (*see also* synthetic polymeric
  biomaterials)
 tissue engineering
  bone regeneration 29–30
  cardiovascular tissues 31
  skin regeneration 30–31
polymeric nano-scaffolds 135
polymers of urethanes (PU) 30
polymethylmethacrylate (PMMA) 6, 166
poly(ortho-esters) (POE) 26–27
polysaccharides
 alginate 23–24
 chitin/chitosan 23
 hyaluronan 24–25
polysaccharide scaffolds 135
polyvinylidene difluroride (PVDF) 6, 89

Porter, A. E. 182, 183
powder metallurgy 74, 75
 additive manufacturing 121
 conventional press-and-sinter process
  116–117
 HIP 117, 118
 microwave sintering 119, 120
 MIM 119, 120
 SHS 117–119
 SPS 119
proteins
 collagen 21–22
 fibronectin 22–23
 silk 22
PSD *see* particle size distribution
  analysis (PSD)
PSOC *see* partial state of charge (PSOC)
p-type N/ZnO (nitrogen-doped) photocatalyst
 characterization 207
 chemicals 206
 dye decomposition 208
 dyes used 207
 EDX 210
 FE-SEM 209, 210
 hydrogen production 208
 N/ZnO
  commercial ZnO samples 213
  dye decomposition 211–213
  hydrogen generation 213–214
 PL spectroscopy 211
 preparation 207
 PSD analysis 210–211
 UV-visible absorption spectroscopy 208
 XRD 208, 209
pulsed laser deposition (PLD) 53
PU *see* polymers of urethanes (PU)
PVDF *see* polyvinylidene difluroride (PVDF)

Qiu, Q. 179
quenching technique 51

radial distribution function (RDF) 43
rapid slurry formation (RSF) process 128–129
reciprocating wear 72
renewable technologies 98
response surface methodology (RSM) 195
rheology
 definition 124
 Newtonian fluids 124
 non-Newtonian fluids 124–126

Salehi, M. 142
Sarkar, S. D. 142
SBF *see* simulated body fluid (SBF)
Scuderi, V. 194
self-lubricating material (SLM) 66
semiconductor photocatalyst 194

224   Index

semi-solid metal working 123
    average grain size 130
    viscocity 129–130
separators 87
sericin-based biomaterial 144–145
shear thickening fluids 125
short-range order (SRO) 38
silk 22
simulated body fluid (SBF) 183
sintering process 74
SLM *see* self-lubricating material (SLM)
small-scale installations 100
smart windows 1–2
Sn-15Pb alloy 126, 127
Soballe (1996) 177
SOC *see* state of charge (SOC)
sol-gel method 194–195
solid electrolyte interphase (SEI) 88
solid lubricants 76
solid polymer electrolytes 94–95
solid polymers 6
solid-state electrolytes 84
spark plasma sintering (SPS) 119
Spencer, R. F. 123
SRO *see* short-range order (SRO)
state of charge (SOC) 101
stress shielding 164
superalloys
    Co-based superalloys 114–115
    definition 111
    fabrication (*see* fabrication)
    Fe-Ni-based superalloys 114
    Ni-based superalloys 112–113
    role of elements 113, 114
    wrought superalloys 112
supercapacitors 99
surface modification 66
Svensson, J. S. E. M. 10
synthetic polymeric biomaterials
    alkylene oxalates 28
    monomer unit 29
    PGA 28
    PLA 28
    POE 26–27
    polyacrylates 26, 27
    polyesters 26
    types 26
synthetic polymers 20

tailor-made materials 41
Tanaka, K. 43
TCP *see* tri-calcium phosphate (TCP)
technological evolution 103
TEM *see* transmission electron microscopy
    (TEM)
Tercero, J. E. 175
tetracalcium phosphate (TTCP) 169–170

tetrahedral-bonded amorphous semiconductors
    (TAS) 45
thermal evaporation 51–52
three-dimensional hydrogel 140
3D printing 121
TiO_2/CuO nanocomposite
    absorption spectra 196
    ANOVA 195, 200
    characteristics 195–198
    energy-dispersive X-ray spectrum 197
    experimental design 195
    FE-SEM image 197
    hydrogen generation 202
    methylene blue 199, 200, 201
    optimization study 198, 200, 201
    photocatalytic activity measurements 195
    preparation 194
    synthesis 194–195
    XRD spectra 196
tissue engineering
    biomaterials 20, 136–139 (*see also* polymeric
        biomaterials)
    bone regeneration 29–30
    CaP 148–150
    cardiovascular tissues 31
    chemical hydrogels 140–141
    chitosan 141–142
    fibroin-based biomaterial 143–144
    HA 150
    hydrogels 139
    vs. medical science *vs.* research 20
    nanotechnology 134
    physical hydrogels 139–140
    polyesters 145–148
    sericin-based biomaterial 144–145
    silk 143
    skin regeneration 30–31
Tomaszek, R. 176
topological close-packed phases 114
Touzain, P. 87
transmission electron microscopy (TEM)
    197, 198
tribo wear test machines 71
tri-calcium phosphate (TCP) 169
TTCP *see* tetracalcium phosphate (TTCP)
tungsten trioxide 10
two-fold coordinated amorphous
    semiconductors 45

ultra-high molecular weight polyethylene
    (UHMWPE) 166
uphill process 54
UV-visible absorption spectroscopy 208

vascular endothelial growth factor (VEGF) 31
viologen/Prussian blue compounds 7
Vogel Tammann Fulcher (VTF) model 98

Index

Wakai, F. 172
Wang, C. M. 128
wear mechanism
    erosive 72
    high-stress wear test 73
    low-stress wear test 73
    measurement techniques 71–72
    parameters, affecting 69, 70
    pin disc 73
    reciprocating 72
    tribo wear test machines 71
    types 70
Wolff's law 178
wonder molecule 141

wrought superalloys 112
Wu, Z-J. 175

xerography 59
Xia, X. H. 10
X-ray diffraction (XRD) 208–209
X-ray photoluminescence (PL) spectroscopy 211

Yamada, K. 182

Zhao, Y. 119
Zheng, L. 127
zinc nitrate hexahydrate decomposition *see* p-type
    N/ZnO (nitrogen-doped) photocatalyst